Waste into Weapons

During the Second World War, the United Kingdom faced severe shortages of many essential raw materials. To keep its armaments factories running, the British government enlisted millions of people in efforts to recycle a wide range of materials for use in munitions production. Recycling not only supplied British factories with much-needed raw materials but it also played a key role in the efforts of the British government to maintain the morale of its citizens, to secure billions of dollars in Lend-Lease aid from the United States, and even to uncover foreign intelligence. However, Britain's wartime recycling campaign came at a cost: it consumed many items that would never have been destroyed under normal circumstances, including significant parts of the nation's cultural heritage. Based on extensive archival research, Peter Thorsheim examines the relationship between armaments production, civil liberties, cultural preservation, and diplomacy, making *Waste into Weapons* the first in-depth history of twentieth-century recycling in Britain.

Peter Thorsheim is Professor of History at the University of North Carolina at Charlotte.

Dedicated to my parents,
Howard and Julie Thorsheim

Studies in Environment and History

Editors

J. R. McNeill, *Georgetown University*
Edmund P. Russell, *University of Kansas*

Editors Emeritus

Alfred W. Crosby, *University of Texas at Austin*
Donald Worster, *University of Kansas*

Other Books in the Series

Kieko Matteson *Forests in Revolutionary France: Conservation, Community, and Conflict, 1669–1848*
George Colpitts *Pemmican Empire: Food, Trade, and the Last Bison Hunts in the North American Plains, 1780–1882*
Micah Muscolino *The Ecology of War in China: Henan Province, the Yellow River, and Beyond, 1938–1950*
John Brooke *Climate Change and the Course of Global History: A Rough Journey*
Emmanuel Kreike *Environmental Infrastructure in African History: Examining the Myth of Natural Resource Management*
Paul Josephson, Nicolai Dronin, Ruben Mnatsakanian, Aleh Cherp, Dmitry Efremenko, and Vladislav Larin *An Environmental History of Russia*
Gregory T. Cushman *Guano and the Opening of the Pacific World: A Global Ecological History*
Sam White *Climate of Rebellion in the Early Modern Ottoman Empire*
Alan Mikhail *Nature and Empire in Ottoman Egypt: An Environmental History*
Edmund Russell *Evolutionary History: Uniting History and Biology to Understand Life on Earth*
Richard W. Judd *The Untilled Garden: Natural History and the Spirit of Conservation in America, 1740–1840*
James L. A. Webb, Jr. *Humanity's Burden: A Global History of Malaria*
Frank Uekoetter *The Green and the Brown: A History of Conservation in Nazi Germany*
Myrna I. Santiago *The Ecology of Oil: Environment, Labor, and the Mexican Revolution, 1900–1938*
Matthew D. Evenden *Fish versus Power: An Environmental History of the Fraser River*
Nancy J. Jacobs *Environment, Power, and Injustice: A South African History*
Adam Rome *The Bulldozer in the Countryside: Suburban Sprawl and the Rise of American Environmentalism*
Judith Shapiro *Mao's War against Nature: Politics and the Environment in Revolutionary China*
Edmund Russell *War and Nature: Fighting Humans and Insects with Chemicals from World War I to Silent Spring*
Andrew Isenberg *The Destruction of the Bison: An Environmental History*
Thomas Dunlap *Nature and the English Diaspora*
Robert B. Marks *Tigers, Rice, Silk, and Silt: Environment and Economy in Late Imperial South China*

Continued after the Index

Waste into Weapons

Recycling in Britain during the Second World War

PETER THORSHEIM
University of North Carolina, Charlotte

CAMBRIDGE
UNIVERSITY PRESS

32 Avenue of the Americas, New York NY 10013-2473, USA

Cambridge University Press is part of the University of Cambridge.

It furthers the University's mission by disseminating knowledge in the pursuit of education, learning and research at the highest international levels of excellence.

www.cambridge.org
Information on this title: www.cambridge.org/9781107492097

First published 2015
First paperback edition 2016

A catalogue record for this publication is available from the British Library

Library of Congress Cataloguing in Publication data
Thorsheim, Peter.
Waste into weapons : recycling in Britain during the Second World War / Peter Thorsheim (University of North Carolina, Charlotte).
pages cm. – (Studies in environment and history)
ISBN 978-1-107-09935-7 (Hardback)
1. World War, 1939–1945–Great Britain–Equipment and supplies. 2. Defense industries–Great Britain–History–20th century. 3. Recycling (Waste, etc.)–Great Britain–History–20th century. 4. Salvage (Waste, etc.)–Great Britain–History–20th century. 5. Scrap metals–Recycling–Great Britain–History–20th century. 6. Cultural property–Great Britain–History–20th century. 7. Civil rights–Great Britain–History–20th century. 8. Great Britain–History, Military–20th century. 9. Great Britain–Foreign relations–1936–1945. 10. Great Britain–Social conditions–20th century. I. Title.
D759.T47 2015
363.72'82094109044–dc23 2015008863

ISBN 978-1-107-09935-7 Hardback
ISBN 978-1-107-49209-7 Paperback

Contents

Figures

Tables

Acknowledgments

I am deeply grateful for the support that many gave to me during the long process of researching and writing this book. To the family members, friends, mentors, colleagues, students, archivists, librarians, editors, and others who made my work both possible and enjoyable, I offer my heartfelt thanks.

Back in the late 1980s, in my first term as an undergraduate at Carleton College, I had the great fortune to take part in Robert Bonner's seminar on war and society. In the years that followed, he introduced me to the historical study not just of war, but also of the environment, of the British Isles, and of the writing of history. Bob inspired me to become a historian, and this book owes much to the questions that he asked in his incomparable classes.

I wish to acknowledge the essential financial support that I received in the form of a Bernadotte Schmitt Grant from the American Historical Association, a summer stipend from the National Endowment for the Humanities, and, from the University of North Carolina at Charlotte, funds from the College of Liberal Arts and Sciences' Small Grants program, the Faculty Research Grants program, and the Office of International Studies. Thanks are also due to my colleagues at UNC Charlotte for granting me leaves from teaching during which I could research and write this book.

For their kind permission to reproduce images or quote from unpublished materials, I wish to thank the Bruce Castle Museum, Lord Addison, the Cabinet Office Historical and Records Section, the Churchill Archives Centre, Her Majesty Queen Elizabeth II, Getty Images, Lord Halifax, June Hopkins, the Imperial War Museum, Evelyn Jackson, London

Metropolitan Archives, the Trustees of the Harold Macmillan Book Trust, the Modern Records Centre, the National Library of Wales, Royal Voluntary Services, the Parliamentary Archives, UNISON, the Walsall Local History Centre, and Westminster City Archives.

For their camaraderie, encouragement, and generous hospitality, I am grateful to everyone who has made me feel welcome during my travels in Britain, especially Catherine Avent, Vincent Hall, Philip Holland, Evelyn Jackson, Gerry Kearns, Stephen Lowther, Bill Luckin, Hilary Marland, Rosemarie Nief, Jonathan Reinarz, Sebastian van Strien, Pete Walton, and Philip Woods.

My thoughts on the history of recycling have been enriched immeasurably by the opportunity to present my research and interpretations at Cambridge University, Davidson College, Kingston University, the Rachel Carson Center, and UNC Charlotte. In addition, I greatly appreciate the opportunities that I have had to exchange ideas with colleagues at meetings organized by the Anglo-American Conference of Historians, the North American Conference for British Studies, the American Society for Environmental History, the European Society for Environmental History, and at a workshop on the environmental history of the Second World War held in Helsinki.

Some of the material presented in this book, especially parts of Chapter 8, appeared previously in Peter Thorsheim, "Salvage and Destruction: The Recycling of Books and Manuscripts in Great Britain during the Second World War," *Contemporary European History* 22, no. 3 (2013): 431–52. I am grateful to Heike Weber and Ruth Oldenziel, the guest editors of the special issue in which that article appeared; Josie McLellan, coeditor of the journal; and Sue Tuck, journals production manager; for their insightful comments and suggestions, and for permitting me to make use of a revised version of that article here.

Special thanks go to the following staff members at UNC Charlotte: Peggy Hoon for her guidance with copyright and legal issues, Ceily Hamilton and Brian Mosley for solving glitches involving computing and scanning, Ann Davis for filling my many ILL requests efficiently, and Leigh Robbins and Linda Smith for keeping the History Department running well every day.

Among the many terrific colleagues at UNCC who supported my work on this book, I wish to single out two good friends for their exceptional help. John David Smith provided sound and steady counsel, read my manuscript meticulously, and offered invaluable advice about publishing. The late Jim McGavran, who shared my love of nature, of writing, and of archives, generously read my manuscript and offered many insightful suggestions only weeks before the tragic stroke that ended his life.

Throughout the editorial process, I benefited greatly from the ideas and expertise of Jeethu Abraham, Dana Bricken, Deborah Gershenowitz, Patterson Lamb, John McNeill, Edmund Russell, Elizabeth Shand, Holly Johnson, and the two anonymous reviewers who read the manuscript for Cambridge University Press.

In addition to those already named, I am extremely grateful for the support and friendship of Andy Boram, Jim Clifford, Mark Crowley, Sandra Dawson, Helen Ford, Jacob Hamblin, Christopher Hamlin, Lyman Johnson, Simo Laakkonen, Christof Mauch, Martin Melosi, Heather Perry, Steve Sabol, Djahane Salehabadi, Susan Strasser, Joel Tarr, Richard Tucker, Timo Vuorisalo, and Mark Wilson.

Words are inadequate to express how much I appreciate my wife, Gina Campbell, and our sons Erik and Jacob, for their love, support, sacrifices, and patience – especially when this project required my absence from them. Finally, I wish to acknowledge the two people who first ignited in me a love of learning, a concern for the environment, and an awareness of the tragedy of war: my parents Howard and Julie Thorsheim. I dedicate this book in their honor.

Introduction

> In this war the very dustbin has become our ally. Nothing is waste, nothing rubbish.
>
> –James Marchant, 1943[1]

The Second World War touched the lives of virtually everyone on the planet and caused the deaths of approximately fifty million individuals – more than half of them civilians. The fighting stretched from deserts to the Arctic, and from remote forests to many of the world's largest cities. Beyond its incomprehensible toll in terms of human suffering, the war consumed vast amounts of money and resources, and it destroyed countless historic buildings, manuscripts, and works of art.

Paradoxically, although the Second World War was the most devastating conflict in history, it inspired people to conserve and recycle resources to an extent never seen before or since. At the same time that the warring nations tried to inflict maximum destruction on their enemies, they sought to use raw materials as efficiently as possible – a phenomenon that led one British newspaper headline writer to proclaim, nonsensically, that there was to be "No Waste in War-Time."[2] One way that governments reduced waste was by rationing. Although rationing allowed goods to stretch further and be distributed more equitably, it did not, of course, increase the overall supply of raw materials. To feed the war's voracious appetite for munitions, many countries turned to salvage: the recycling of waste and "useless" articles. Salvage could not guarantee any nation's

[1] James Marchant, *World Waste and the Atlantic Charter* (Oxford: Blackwell, 1943), 14.
[2] Percy A. Harris, letter to the editor, *Times* (London), 25 Sept. 1939, 9.

victory, but many wartime leaders believed that a failure to make use of scrap would ensure defeat.

Between 1939 and 1945, salvage played a crucial role in the United Kingdom, a country that depended greatly on imports to supply it with the food and raw materials that it consumed. The war increased the cost of many imports and made others unobtainable at any price. Even when foreign supplies could be found, Britain lacked sufficient ships, sailors, and foreign currency to import everything it needed. The situation only grew worse as German submarines and ships sank a substantial fraction of Britain's merchant fleet.[3] Domestic recycling avoided all of these pitfalls.

Officials used both coercion and persuasion to promote salvage during the war. Through a series of regulations, the British government made it a crime to discard or destroy ferrous and nonferrous metals, paper, rags, string, rubber, animal bones, and food scraps. In addition to requiring people to recycle waste, officials instituted salvage drives through which they tried to convince members of the public to contribute items that they never would have discarded in peacetime, such as antiques, old coins, books, manuscripts, and even brand new pots and pans. If people were not sufficiently generous, officials warned, they would use the wartime powers Parliament had granted them to seize private property. Inverting the biblical injunction to beat swords into plowshares, the British wartime state called on farmers to hand in old agricultural implements so they could be recycled into guns and bombs, the swords of twentieth-century combat.[4] Government ministries, working with local authorities and volunteers, conducted surveys, engaged in propaganda campaigns, and finally, resorted to compulsion in an effort to supply Britain's factories with raw materials.

Despite the centrality of salvage in Britain during the Second World War, few studies have examined it in any detail. Floud and McCloskey's classic survey of British economic history ignores the subject altogether, and David Edgerton's 2011 book, *Britain's War Machine: Weapons, Resources, and Experts in the Second World War*, devotes only a single sentence to it.[5] The first historians to pay significant attention to recycling in wartime were Angus Calder and Norman Longmate, whose

[3] Ministry of Supply, "Memorandum on British Methods of Raw Material Conservation," Sept. 1943, 2, National Archives of England, Wales, and the United Kingdom, Kew (hereafter NA), AVIA 46/490.

[4] "Shells from Ploughshares," *News Chronicle*, 9 Mar. 1940.

[5] See Roderick Floud and Deirdre McCloskey, *The Economic History of Britain since 1700, Vol. 3: 1939–1992*, 2d ed. (Cambridge: Cambridge University Press, 1994); David

panoramic histories of the British home front appeared more than four decades ago. Despite their detailed discussion of the subject, Calder and Longmate approached salvage almost entirely from the perspective of its role in mobilizing civilians into the war effort.[6] Only recently has the history of waste and recycling emerged as a subject in its own right. Influenced by the groundbreaking books of Martin Melosi and Susan Strasser on the history of waste disposal in the United States, researchers have begun to explore these topics in many other countries, including the United Kingdom.[7] In recent years, historians have devoted a growing amount of attention to connections between war and the environment, and this development has done much to influence the study of resource use and recycling.[8]

As the environmental historian Tim Cooper has observed, the two world wars caused a "rediscovery of recycling" in Britain. He demonstrates that the exigencies of war led experts to study the composition of municipal waste and to seek ways to extract value from something they had previously considered worthless. The notion of "'waste' as a physical material with its own characteristics and potential uses," he argues, "was a wartime discovery."[9] Cooper posits that wartime shortages led most people in Britain to temporarily abandon what Bill Luckin

Edgerton, *Britain's War Machine: Weapons, Resources, and Experts in the Second World War* (Oxford: Oxford University Press, 2011), 205.

6 Angus Calder, *The People's War: Britain, 1939–1945* (New York: Pantheon, 1969); Norman Longmate, *How We Lived Then: A History of Everyday Life during the Second World War* (London: Hutchinson, 1971).

7 Martin V. Melosi, *Garbage in the Cities: Refuse, Reform, and the Environment, 1880–1980* (College Station: Texas A&M University Press, 1981); Susan Strasser, *Waste and Want: A Social History of Trash* (New York: Metropolitan Books, 1999); Bill Luckin, "Pollution in the City," in *The Cambridge Urban History of Britain, Vol. 3: 1840–1950*, ed. Martin Daunton (Cambridge: Cambridge University Press, 2000), 207–28; Peter Hounsell, *London's Rubbish: Two Centuries of Dirt, Dust and Disease in the Metropolis* (Stroud: Amberley, 2013).

8 Carl A. Zimring, *Cash for Your Trash: Scrap Recycling in America* (New Brunswick, N.J.: Rutgers University Press, 2005); Hugh T. Rockoff, "Keep on Scrapping: The Salvage Drives of World War II" (Sept. 2007), NBER Working Paper Series, vol. w13418, 2007, http://ssrn.com/abstract=1014795, accessed 17 Dec. 2012. In Nazi-occupied Europe, the recycling of cultural artifacts was often an attempt to purge national symbols from public space. See Elizabeth Campbell Karlsgodt, "Recycling French Heroes: The Destruction of Bronze Statues under the Vichy Regime," *French Historical Studies* 29, no. 1 (2006): 143–81; Chad Denton, "'Récupérez!' The German Origins of French Wartime Salvage Drives, 1939–1945," *Contemporary European History* 22, no. 3 (2013): 399–430.

9 Tim Cooper, "Challenging the 'Refuse Revolution': War, Waste and the Rediscovery of Recycling, 1900–50," *Historical Research* 81, no. 214 (2008): 710–31.

has termed "the refuse revolution" of the late nineteenth and early twentieth centuries. This revolution, Luckin suggests, involved a major paradigm shift in how British citizens understood rubbish. Previously, many people viewed household discards as valuable materials that should not be allowed to go to waste; after the revolution, they considered rubbish both worthless and harmful. Instead of reusing or recycling waste, they sought to destroy it as quickly and completely as possible, through incineration, ocean dumping, or land burial.[10]

In contrast with Cooper's focus on the influence that local and national officials exerted in developing wartime recycling policies, the historical geographer Mark Riley examines changes in popular attitudes toward waste disposal during the war.[11] Both Riley and Cooper suggest that during the Second World War many Britons abandoned – at least temporarily – the idea that refuse was worthless and instead considered it a source of valuable materials. This observation is extremely significant, but the revaluing of rubbish was not the only conceptual change that facilitated Britain's wartime salvage program. Equally important was a revolution in the very definition of waste.

Recycling is usually thought of as an activity that derives value from materials that are broken or at the end of their useful life. In wartime, however, many of the objects that people recycled were still in use. To maximize the amount of raw material available for war production, Britain's leaders sought to redefine all "unnecessary" items as devoid of value. Once deemed useless, these "wastes" could be recycled into weapons or other necessities. Using a form of doublespeak that would later become familiar to readers of Orwell's *1984*, the British government suggested that to preserve articles was to waste them. In the name of preventing such things from "going to waste," they were destroyed (salvaged) so they could be turned to a more "productive" use as tools of destruction. Attentive to these contradictions, the archivist Hilary Jenkinson observed in 1944 that "the word salvage . . . has in current usage the initial disadvantage of two exactly opposite meanings: that of 'salving' for the national need papers of all sorts, from omnibus tickets

[10] Cooper, "Challenging," 719; Luckin, "Pollution in the City;" and Tim Cooper, "Burying the 'Refuse Revolution': The Rise of Controlled Tipping in Britain, 1920–1960," *Environment and Planning A* 42, no. 5 (2010): 1033–48.

[11] Mark Riley, "From Salvage to Recycling: New Agendas or Same Old Rubbish?" *Area* 40, no. 1 (2008): 79–89, esp. 80.

(literally) to records, which are presumed to be valueless; and that of saving papers which are not valueless from 'salvage.'"[12]

The massive destruction wrought by war lessened inhibitions against the annihilation not only of human life but also of the natural and built environment.[13] One consequence of this violent mindset was the ease with which many soldiers and civilians rationalized the destruction of cultural artifacts in wartime. Such feelings seemed to justify not only the incineration of enemy cities and their inhabitants but also a callousness toward people, places, and artifacts closer to home. Although some considered the extraordinary reach of salvage activities to be a regrettable necessity, others – perhaps influenced by the anti-historicism that the Italian writer F. T. Marinetti had expressed in his 1909 Futurist Manifesto – welcomed the opportunity to destroy old books and manuscripts and thereby lessen the burden that the past exerted on the present.[14] Some contemporaries hinted that the urge to destroy was deeply imbedded within the human psyche. An unsigned article in the *Times* in 1940 noted that a "demon of destructiveness . . . lurks within every one of us. . . . The mildest of paper-throwers will sooner or later find this noble rage attack him. In that moment everything will go into one gorgeous heap on the floor – old manuscripts, over which treasures of painstaking were once spent; old newspaper cuttings once deemed, goodness knows how erroneously, to have been of interest; old executorship accounts of uncles and aunts long since wound up." The thrill of throwing things into the salvage collection was not limited to paper.

[12] Hilary Jenkinson, "British Archives and the War," *American Archivist* 7, no. 1 (1944): 1–17, quotation from 12. For an account of the man by one of his closest colleagues, see Roger H. Ellis, "Recollections of Sir Hilary Jenkinson," *Journal of the Society of Archivists* 4, no. 4 (1971): 261–75. For an insightful analysis of the often blurry distinction between books as texts and as tangible objects, see Leah Price, *How to Do Things with Books in Victorian Britain* (Princeton, N.J.: Princeton University Press, 2012). Garbage and recycling have recently gained the attention of philosophers as well. For a notable example of such work, and one which includes particularly interesting arguments about the centrality of ideas about the passage of time, see John Scanlon, *On Garbage* (London: Reaktion Books, 2005).

[13] For two insightful explorations of these ideas, see Modris Eksteins, *Rites of Spring: The Great War and the Birth of the Modern Age* (Boston: Houghton Mifflin, 1989); Edmund Russell, *War and Nature: Fighting Humans and Insects with Chemicals from World War I to Silent Spring* (Cambridge: Cambridge University Press, 2001).

[14] F. T. Marinetti, "The Founding and Manifesto of Futurism," in R. W. Flint, ed., *Marinetti: Selected Writings*, trans. R. W. Flint and Arthur A. Coppotelli (London: Secker & Warburg, 1972), 39–44. The Futurists' embrace of speed, technology, and violence led many of them to welcome the outbreak of war in 1914 as a catalyst for revolutionary change.

According to the same observer, people could experience "the same fierce joy" in "hurling away saucepans and bedsteads, in uprooting railings, in dismembering bicycles."[15]

Much of the ferrous metal that the British government salvaged in wartime came from the gates and railings that surrounded many of the country's parks, schools, churches, government buildings, and houses.[16] In total, railings contributed over half a million tons of metal to the war effort. Officials declared that these railings were nothing but unnecessary scrap littering the landscape, although many of their owners sharply disagreed.

Another major source of raw material that officials would never have seized in peacetime was "blitz scrap," a term that referred to the full range of materials that could be salvaged from buildings damaged or destroyed by German bombs. Some blitz scrap was used to make emergency repairs to buildings, but much of it was transformed into bombs and artillery shells. In an extraordinary demonstration of wartime zeal, officials often chose to demolish rather than repair bomb-damaged buildings simply to harvest their metal. This added considerably to the war's toll on architecture, and like the recycling of manuscripts, it further depleted Britain's cultural inheritance. The authorities even found an important use for millions of tons of pulverized brick and stone, which provided construction material for airfields.

Salvage involved not only a physical transformation but often a symbolic one as well. In the case of both waste and reclaimed items, the government had to convince people of the paradoxical notion that many types of goods were simultaneously useless to them as individuals and essential to the nation's survival. To persuade people to part with their possessions and, equally as important, to justify why the government should be allowed to take them without paying significant compensation, the authorities asserted that these items had no cultural, historical, or sentimental significance. A book was only so much waste paper; a coin collection was simply scrap metal. Their real value resided in their potential to be turned into munitions. Many people eagerly embraced this idea of metamorphosis, and some went so far as to ask that their contributions be used for the manufacture of a particular type of weapon. Recognizing that many of the items people donated through municipal collections or

[15] "Throwing Away," *Times* (London), 20 July 1940, 5.
[16] "Railings Removal," *Public Cleansing and Salvage*, Dec. 1942, 123.

sold to merchants possessed a monetary value that exceeded their worth as scrap, officials emphasized repeatedly that strenuous efforts were required "to prevent any form of Black Market activity and to ensure that everything given in should be disposed of correctly."[17] A failure to diminish the resale value of medals, door knockers, and statues might well tempt merchants to find a buyer who would pay more for them than the government-controlled scrap price. To reduce the chance that items would be diverted from the salvage stream, officials promoted actions that transformed unique artifacts into undifferentiated raw materials. Saucepans and statues were smashed, railings were cut up, and papers were shredded.

Conscious of how its recycling policy would influence attitudes both at home and abroad, the British government used salvage as a tool to sustain the morale of British civilians and to persuade the United States to provide the aid that the United Kingdom so desperately needed. American officials paid close attention to all aspects of the British economy during the war, and they made it clear that they considered salvage a crucial demonstration of Britain's commitment to winning the war. In addition to serving material, economic, psychological, and diplomatic roles, salvage drives aided British intelligence-gathering efforts by helping to uncover foreign business directories, maps, and other documents.

Wartime recycling dissolved the boundaries between public and private life, paid and unpaid labor, soldiers and civilians, munitions and everyday objects, and home and battlefield. It also challenged gender divisions. Despite the many new opportunities and demands that women experienced during the Second World War, patriarchy remained strong, and ideas about the home as a feminine realm continued to exert a major influence. Although officials encouraged the belief that household recycling was a logical extension of traditional feminine roles, women's participation in wartime salvage work subverted gender norms by encouraging people to think of women as combatants and the domestic sphere as part of the war machine.[18] It infused one of the most mundane of household tasks – taking out the rubbish – with life and death meaning.

[17] Notes of a meeting at the Directorate of Salvage and Recovery on 17 Aug. 1942, NA, AVIA 22/3088.

[18] For an insightful discussion of these issues, see Cynthia H. Enloe, *Maneuvers: The International Politics of Militarizing Women's Lives* (Berkeley: University of California Press, 2000), and Tammy M. Proctor, "'Patriotism Is Not Enough': Women, Citizenship, and the First World War," *Journal of Women's History* 17, no. 2 (2005): 169–76.

In the process, every home became a source of war materiel, and every man, woman, and child became a de facto munitions worker. Salvage epitomized what Miriam Cooke has called the "militarization of daily life."[19] So great was the blurring between the civil and military realms that some pacifists refused on moral grounds to recycle their waste paper – an act of civil disobedience that was a criminal offense in wartime. For although the British state excused conscientious objectors from both military service and work in munitions factories during the war, it provided no similar exemption if civilians, male or female, refused to participate in the government's efforts to recycle their household waste into weapons.[20]

To a greater extent than any other wartime activity in Britain, salvage exemplified total war. It provided both a material and a symbolic link between civilians, their government, and the war as a whole. Salvage thus served as a vital focus for people's efforts to make sense of the war and to understand their role in it. It also acted as a barometer of swiftly changing military and strategic developments. As one British official observed,

> Until the German occupation of the Low Countries and France in the summer of 1940, the supply position of zinc was good and its use was therefore encouraged as a substituting material for both copper and aluminium. After this period, zinc itself became a material for which substitutes were sought. Similarly tin, and more frequently lead, were used as substituting materials until the Japanese hold on Burma and Malaya, and, in the case of lead, the uncertainty of shipments from Australia, completely altered the supply situation.[21]

In response to the volatile supply situation, the British government organized a series of high-profile campaigns to collect aluminum (July 1940), iron railings (September 1941), rubber (February 1942), nonferrous metals (October 1942), and paper and books (multiple times).

Britain's wartime recycling experience was rife with contradiction. Far from boosting morale, salvage often had the opposite effect when the authorities' fervent desire to maximize salvage led to infringements of civil liberties and property rights and to the destruction of historical and cultural artifacts. Thousands of people objected to the requisition of their property, and many denounced as insulting the meager levels of

[19] Miriam Cooke, "War, Gender, and Military Studies," *NWSA Journal* 13, no. 3 (2001): 181–8, esp. 186.

[20] "Hastings Prosecution," *Public Cleansing and Salvage*, Nov. 1943, 128.

[21] Ministry of Supply, "Memorandum on British Methods of Raw Material Conservation," Sept. 1943, 5, NA, AVIA 46/490.

compensation that the government offered. To add insult to injury, some of the materials that people diligently contributed to salvage drives still lay in dumps at the war's end, unused and unwanted. Finally, Britain's wartime recycling campaigns furthered the immense destruction of cultural and historical artifacts that took place during the Second World War. For despite its ostensible goal of preventing waste, the ultimate aim and result of wartime salvage was destructive.

PART I

BEATING PLOWSHARES INTO SWORDS

1

Salvage in Times of Peace and War

> Salvage-work is important at any time; in war-time it is of fundamental importance, especially to us as an island people. We must have greatly increased supplies of raw materials, many of which are normally imported, at a time when the difficulties of carrying sea-borne cargoes is [*sic*] truly great and hazardous. But we must have them, and we turn to substitutes at home.
>
> –J. C. Dawes, 1942[1]

The word "recycling" did not enter widespread use until the 1960s, but human beings have been salvaging broken tools, old clothing, and scrap materials for centuries. Prior to the emergence of the modern environmental movement, the primary impetus for recycling came not from a desire to conserve nonrenewable resources or prevent pollution but rather from an effort to find inexpensive sources of raw materials. Adaptive reuse played a major role in early modern Europe, and it had important military as well as economic consequences. When Henry VIII destroyed England's monasteries in the late 1530s, for example, his government confiscated many of their bells. Made from an alloy of copper and tin, they provided excellent material for the manufacture of cannon. Over the centuries, foundries have also recycled many other civilian items, such as lead pipes and iron tools, into bullets and cannon balls.[2]

[1] J. C. Dawes, "Making Use of Waste Products," *Journal of the Royal Society of Arts* 90 (15 May 1942): 388–408, quotation from 389.

[2] Donald Woodward, "Swords into Ploughshares: Recycling in Pre-Industrial England," *Economic History Review* 38, no. 2 (1985): 175–91; Pierre Desrochers, "Does the Invisible Hand Have a Green Thumb? Incentives, Linkages, and the Creation of Wealth out of Industrial Waste in Victorian England," *Geographical Journal* 175, no. 1 (2009): 3–16.

By the nineteenth century, industrialization and new patterns of consumption caused a dramatic expansion in the amount of waste people generated. London, the Western world's largest city at that time, became a laboratory for how to cope with this vast quantity of unwanted material. In his classic study, *London Labour and the London Poor* (1851–61), the social investigator William Mayhew described in vivid detail the recycling activities of people whom the Victorians referred to as tatters, pickers, dredgermen, mud-larks, and rag-and-bone men.[3] Such workers played a crucial role not only in keeping nineteenth-century cities from drowning in refuse but also in supplying industry with essential raw materials. Despite the important contributions these scavengers made, many considered them to be disreputable, dirty, and out of place in the modern world. They also inspired numerous cultural responses, ranging from novels such as Dickens's *Our Mutual Friend* in the 1860s to the BBC television sitcom *Steptoe and Son* a century later.

Prejudice against those who made their living from salvaging castoffs intensified as local and national governments assumed greater responsibility and control over the urban environment. By 1900, most Britons no longer saw waste materials and the people who collected them as valuable; instead, they viewed them as sources of disorder and contamination. The overwhelming priority was to remove wastes from populated areas before they could cause offense or illness – not to ensure that they were reused or recycled.[4] As a result, many towns and cities encouraged residents to burn their waste paper and other combustible rubbish to reduce the burden on municipal waste disposal services.

[3] Henry Mayhew, *London Labour and the London Poor*, vol. 1 (London, 1851), 3, 4.

[4] James Winter, *London's Teeming Streets, 1830–1914* (London: Routledge, 1993); Mark Riley, "From Salvage to Recycling: New Agendas or Same Old Rubbish?" *Area* 40, no. 1 (Mar. 2008): 79–89; Tim Cooper, "Modernity and the Politics of Waste in Britain," in *Nature's End: History and the Environment*, ed. Sverker Sörlin and Paul Warde (London: Palgrave Macmillan, 2009), 247–72. For theoretical perspectives on the modernist impulse, see Patrick Joyce, *The Rule of Freedom: Liberalism and the Modern City* (London: Verso, 2003); James C. Scott, *Seeing Like a State: How Certain Schemes to Improve the Human Condition Have Failed* (New Haven, Conn.: Yale University Press, 1999). On scavenging and waste disposal in the United States, see Susan Strasser, *Waste and Want: A Social History of Trash* (New York: Metropolitan Books, 1999); Martin V. Melosi, *The Sanitary City: Urban Infrastructure in America from Colonial Times to the Present* (Baltimore, Md.: Johns Hopkins University Press, 2000); Carl A. Zimring, *Cash for Your Trash: Scrap Recycling in America* (New Brunswick, N.J.: Rutgers University Press, 2005).

THE GREAT WAR

Attitudes toward waste and recycling shifted dramatically between 1914 and 1918 in response the terrible human, financial, and material toll of the First World War. According to one contemporary estimate, the conflict cost the belligerents a total of one billion dollars every week.[5] Out of sheer necessity, people in many countries sought ways to make old or worn-out items useful again. In the face of extreme deprivation and famine on the German home front, women in that country took up recycling on a massive scale.[6] Among the Central Powers, wartime recycling was not limited to wastes but grew to involve the wholesale commandeering of many items that were still in use. As Modris Eksteins has observed of Germany, "Over eighteen thousand church bells and innumerable organ pipes were donated to the war effort, to be melted down and used for arms and ammunition."[7] In Austria-Hungary, the government announced the requisition of all church bells in December 1917.[8]

State-sponsored recycling in Britain began in the army, and it was there that it developed most fully during the Great War. As the monumental offensives of 1916 failed to achieve anything other than an incomprehensible number of casualties, officials looked for new ways to extract the human and material resources needed to continue fighting. Conscription was of course the most notable example, but recycling also played a significant role in promoting what many referred to as national efficiency.[9] The need was acute: artillery shells alone were consuming more than half of Britain's entire output of steel.[10] Over time, the British Army established an elaborate system to salvage a wide range of battle scrap from the Western Front, including shattered guns and equipment,

[5] Isaac F. Marcosson, "The Salvage of War," *Saturday Evening Post*, 5 Jan. 1918, 14.

[6] Heike Weber, "Towards 'Total' Recycling: Women, Waste and Food Waste Recovery in Germany, 1914–1939," *Contemporary European History* 22, no. 3 (2013): 371–97.

[7] Modris Eksteins, *Rites of Spring: The Great War and the Birth of the Modern Age* (Boston: Houghton Mifflin, 1989), 202.

[8] "The Austrian Army," *Times* (London), 10 Dec. 1917, 7.

[9] "Report upon the Work of the Quartermaster-General's Branch of the Staff and Directorates Controlled, British Armies in France and Flanders, 1914–1918," n.d., NA, WO 107/69; "Notes on Salvage," n.d., NA, WO 107/71; "Brief History of the Army Salvage Branch," Oct. 1919, NA, WO 107/73.

[10] W. R. Lysaght, "Production of Steel for the Year 1916," n.d. (copy), Papers of Christopher Addison, Department of Special Collections and Western Manuscripts, Bodleian Libraries, University of Oxford (hereafter Addison Papers), MS. Addison dep. c. 43, folios 75–76. Extracts from the Papers of Christopher Addison are reproduced with the kind permission of William Addison, fourth Viscount Addison.

packing cases, barbed wire (both German and Allied), and used cartridges.[11] The processing of this material was extremely hazardous because the consignments often included undetonated grenades and ammunition. On one day in 1917, for example, a young worker at Woolwich Arsenal in London, identified in the surviving documentation simply as "boy Widdowson," lost four fingers when a cartridge that he was attempting to clean blew up in his hand.[12]

Risky as it was, the refurbishing of used ammunition resulted in enormous cost savings. As the minister of munitions put it in a speech to the House of Commons in the summer of 1917, "We are now able to reform hundreds of thousands of 18-pounder cartridge cases per week. When it is remembered that the price of a new case is about 7s. [shillings], and that it can be reformed four times, and that we are reforming cases at the cost of 4d. [pence] a case, the importance of this branch of work is obvious."[13] During the final year of the war, the British army shipped home 389,000 tons of scrap from the battlefields of France.[14] Operating on a double shift, workers refurbished 750,000 cartridges a week.[15] Salvage also occurred in other war theaters. In 1918, the British began shipping battle scrap from Egypt and Greece to factories in India, where it was processed for reuse.[16]

On 5 January 1918, three days before Woodrow Wilson delivered his Fourteen Points speech to a joint session of Congress, the *Saturday Evening Post* published a major article about the British war effort. Written by the prominent editor and author Isaac F. Marcosson, it was entitled "The Salvage of War." The upbeat message that it put forth was consistent not only with the one that Wilson was about to proclaim but also with the ideas of "scientific management" propounded by the American mechanical engineer Frederick Winslow Taylor. "Out of the vast vortex that to-day engulfs men, money, and materials," wrote Marcosson, "there is coming a tremendous lesson in economy that will

[11] Report by director of salvage to board, 12 Nov. 1917, NA, MUN 4/2232; Department of the Surveyor General of Supply, Salvage Board, minutes, 15 Dec. 1917, NA, MUN 4/2232.
[12] "Proceedings of a Court of Enquiry Held at E31 Royal Arsenal, Woolwich on the 7th Day of December 1917," NA, MUN 4/2232.
[13] *Parliamentary Debates*, Commons, 28 June 1917, vol. 95, col. 575.
[14] J. Stevens, report on salvage operations, 7 Oct. 1918, NA, MUN 4/6606.
[15] Ministry of Munitions, Salvage Department, "Salvage Returned from Overseas," n.d., NA, MUN 4/2232.
[16] J. Stevens, report on salvage operations, 7 Oct. 1918, NA, MUN 4/6606.

make peace more efficient and more orderly." Despite the horrors of the trenches, he insisted, "War is not all waste."[17] As Marcosson described it, the British had developed a system for cleaning up the battlefield characterized by breathtaking efficiency – and lack of decorum. "Almost before the flame and fury of battle subside," he wrote, salvage teams begin "gathering up steel helmets, rifles, belts, haversacks, bayonets, shell cases, unexploded bombs and grenades, clothes, leggings, shoes; in fact, every scrap of stuff that can be transported. . . . Only the dead remain wherethey fall. They alone are the unsalvaged." Marcosson's article did not specify the total cost savings that recycling provided, but he claimed that the reconditioning of clothing and blankets alone saved the British army over $12 million each year.[18]

Eight months after his article appeared, Marcosson wrote to Surveyor General of Supplies Andrew Weir to request information about several topics, among them salvage. This time he wanted to know not only about recycling within the military but also on the home front.[19] Just as they would do during the Second World War, British officials were anxious to convince the American public that Britons were doing everything in their power – including diligent salvage – to bring about victory. Unfortunately, the British government kept few records about the nature or quantity of materials recycled on the home front during the Great War.[20] As a memo from a government statistician explained, most civilian salvage was carried out "through Trade Channels for which no figures are available." He estimated that local government authorities had collected only about £100,000 worth of salvaged materials in 1917 and added that "this is a small fraction of what might be saved. The possibilities of Salvage through Local Authorities is [*sic*] estimated at £5,000,000 at least."[21]

Christopher Addison, a Liberal MP who represented Hoxton, a borough constituency in the East End of London, served as parliamentary secretary to the Ministry of Munitions in 1915 and 1916, and as minister of munitions from December 1916 to July 1917. The ministry was an enormous organization, the headquarters of which employed over

[17] Isaac F. Marcosson, "The Salvage of War," *Saturday Evening Post*, 5 Jan. 1918, 14–16, 97–105. A slightly revised version of this article was republished in Isaac F. Marcosson, *The Business of War* (New York: John Lane, 1918), chpt. 7.

[18] Marcosson, "Salvage of War," 14.

[19] Isaac F. Marcosson to Andrew Weir, 26 Sept. 1918, NA, MUN 4/6606.

[20] James T. Currie to Isaac F. Marcosson, 8 Oct. 1918, NA, MUN 4/6606.

[21] G. Goodwin Self, "British Salvage Operations," 3 Oct. 1918, NA, MUN 4/6606.

9,700 people by March 1917.[22] On 2 January 1917, Addison wrote Field Marshal Douglas Haig that "the salvage of used ammunition . . . vitally affects the output which we shall be able to provide for you during the next six months. . . . In making arrangements for the very large programme which you put before us last September we have had . . . to explore the whole of the world's available resources of metals." Referring to "the very serious prospective shortage of copper," Addison urged Haig to order soldiers to remove the firing bands from used shell cases. "We need to recover every ounce of copper possible," he explained.[23] Following a visit to France a short time later, Addison sent another letter to Haig. In it, he noted that in 1916 the British had salvaged roughly half of the quick-firing cartridge cases and wooden shipping boxes sent to the front. Yet he pointed out that several items had not been recycled to any significant extent, including fuse cylinders, tin boxes, cartridge cylinders, shrapnel cases, and copper bands. "The salving of the two last items," he added, "presents considerable difficulties, though both are of great value to us in view of the shortage of raw materials."[24]

In the summer of 1916, just two weeks after launching their offensive at the Somme, British officials met to discuss the looming shortage of shell steel. To address this shortfall, the head of shell manufacture, Sir Glynn Hamilton West, argued that battle scrap alone would not solve the problem: the armaments industry should look to the home front for materials.[25] At first, such suggestions struck many as extreme. When the chancellor of the exchequer, Andrew Bonar Law, brought up a historical example of such practices, he did so not as a practical suggestion but as metaphor for sacrifice. In a 1917 speech intended to promote the sale of war bonds, he quoted from Thomas Carlyle's famous history of the French Revolution: "Railings are torn up; hammered into pikes: chains themselves shall be welded together, into pikes. The very coffins

[22] "Organization of the Ministry of Munitions, March 1st. 1917," Addison Papers, MS. Addison dep. c. 25, folio 33. In 1937, Addison became Baron Addison, and in 1945 he was named first Viscount Addison. See Kenneth O. Morgan, "Addison, Christopher, first Viscount Addison (1869–1951)," in *Oxford Dictionary of National Biography* (Oxford: Oxford University Press, 2004).

[23] Christopher Addison to Douglas Haig, 2 Jan. 1917 (copy), Addison Papers, MS. Addison dep. c. 35, folios 8–12.

[24] Christopher Addison to Douglas Haig, 13 Mar. 1917 (copy), Addison Papers, MS. Addison dep. c. 34, folios 115–21.

[25] "Minutes of Proceedings at a Conference on British and Allies Requirements of Shell Steel, Pig Iron, &c.," Armament Buildings, London, S.W., Tuesday, July 18th, 1916, Addison Papers, MS. Addison dep. c. 43, folios 185–189.

of the dead are raised; for melting into balls. All Church-bells must down into the furnace to make cannon; all Church-plate into the mint to make money." Bonar Law followed this quote by asking, "Do not those words stir our blood like the sound of an organ in some vast Cathedral? Do they not . . . set before us the soul of a people? And a people in such a spirit is unconquerable." This line was met with cheers, as was his assertion that the British people would not flinch from any sacrifice needed "to hasten the day of victory."[26]

In contrast to the Second World War, when British officials decided to promote domestic recycling almost immediately, during the Great War they waited until more than three years of fighting had elapsed. The campaign to encourage civilians to recycle their wastes was led by officials from the army and the Ministry of Munitions. In January 1918, Lord Derby, secretary of state for war, oversaw the creation of an organization designed to promote recycling on the home front.[27] A short time later, Sir James MacPherson, under-secretary of state for war, announced,

> Steps are being taken now to deal with civil waste in the same way as with Army waste, and for this purpose a National Salvage Council is being constituted under the chairmanship of the Quartermaster-General, and consisting of representatives of all the Departments interested, in order to provide a co-ordinated scheme of salvage which will have most far-reaching results in the days of scarcity that lie before us. We cannot sufficiently impress upon our minds that every ton of raw material saved or salved in this country not only saves the cost of a corresponding ton of new material, but releases more and more tonnage for the importation of foodstuffs which, once consumed, cannot be reproduced.[28]

As in Germany, recycling on the British home front relied heavily on the unpaid labor of volunteers, most of whom were women and children. In April 1918, the *Times* reported that the National Salvage Council was working with the Kensington branch of the British Women's Patriotic League to organize a house-to-house canvas for salvageable items. "It is hoped," noted the author of the article, "that other boroughs will follow the example of Kensington."[29] In addition to enlisting female volunteers,

[26] "Terms of the Loan," *Times* (London), 12 Jan. 1917, 7; E. H. H. Green, "Law, Andrew Bonar (1858–1923)," in *Oxford Dictionary of National Biography* (Oxford: Oxford University Press, 2004).

[27] Report of meeting at the War Office, 24 Jan. 1918, NA, MUN 4/2232; Keith Grieves, "Stanley, Edward George Villiers, Seventeenth Earl of Derby (1865–1948)," in *Oxford Dictionary of National Biography* (Oxford: Oxford University Press, 2004).

[28] *Parliamentary Debates*, Commons, 20 Feb. 1918, vol. 103, cols. 773–4.

[29] "Valuable Rubbish," *Times* (London), 20 Apr. 1918, 3.

the British government encouraged schoolchildren to collect salvage, especially in rural areas.[30]

In April 1918, the National Salvage Council sent to local authorities across the kingdom a pamphlet that called on them to recycle animal bones, food scraps, metal, rags, and paper. To encourage dustmen to keep salvageable materials separate from useless rubbish, the group encouraged localities to pay them a bonus for the work.[31] It also established a membership organization called the Salvage Club, complete with its own magazine.[32] Building on the salvage of food scraps that had reached a high level of efficiency in the army, the National Salvage Council hoped that civilian kitchens might provide food for municipally run pig farms.[33] Chester had 68 pigs under its care, Liverpool had 200, and the City of London boasted a farm of 1,000 swine.[34]

Reporting on the work of the new organization, the *Times* declared that "salvage is a matter of vital importance to the nation at the present time, and one in which every man, woman, and child in the country can lend assistance."[35] A short time later the newspaper printed a short article, most likely based on a press release, titled "The Importance of Saving Bones": "One of the most important activities of the newly-formed National Salvage Council will concern the use of refuse from private houses. . . . [T]his is an opportunity to help the nation, at no expense – indeed, at a small profit – and with very little trouble, simply by saving up fragments that are usually thrown away or burned. Bones should never be burned or thrown into the dustbin, even after making soup." Instead, they could be made into fertilizer, chicken feed, or even explosives.[36]

In 1918, faced with a growing shortage of paper and cardboard, the British government urged local authorities throughout the country to turn over all documents and books that could "without any harm be destroyed at once."[37] Well before this request, however, many officials had begun to salvage such materials for the war effort. In July 1916, for example, municipal employees in the London borough of Paddington decided to

[30] "Value of Refuse," *Times* (London), 29 Apr. 1918, 3.
[31] "Value of Refuse," *Times* (London), 29 Apr. 1918, 3.
[32] "The Salvage Club," *Times* (London), 2 May 1918, 3.
[33] "Farmers' Option in Cattle Sales," *Times* (London), 16 Mar. 1918, 7.
[34] "Municipal War Work," *Times* (London), 12 Sept. 1918, 3.
[35] "Saving of Waste," *Times* (London), 2 Mar. 1918, 3.
[36] "Importance of Saving Bones," *Times* (London), 19 Mar. 1918, 3.
[37] H. W. S. Francis, memorandum, 20 Mar. 1918, NA, HLG 102/93.

clear out their old records to supply the nation with waste paper. Among the documents they sacrificed were the registration book of a girls' school that contained entries from as early as 1763, local records from the Victorian period on vagrancy, medical relief, and juvenile justice, and an intriguingly titled document that discussed the "allowance of beer and mineral water to nurses." All went to the paper mills for pulping.[38] Officials in Britain also called on communities to give up heavy iron objects to be melted down and made into munitions. This proposal sparked considerable interest, but it brought forth relatively little metal before the war ended. In Edinburgh, city officials removed the railings that surrounded several parks, and in Westminster, the city council offered 150 tons of cast-iron bollards.[39]

In the spring of 1918, the National Salvage Council began holding meetings with local officials to encourage them to recover useful materials from the waste that they collected and to help them find appropriate markets for it.[40] At one of these meetings, the lord mayor of Birmingham stated that his city annually collected 600 tons of old tin cans, and that the tin which came from them sold for £300 per ton. Multiplying the two figures, the headline writers for the *Times* proclaimed "£180,000 a Year from Old Tins."[41] This was totally inaccurate, however. Despite their name, tin cans consisted overwhelmingly of steel, covered by a thin coating of tin. The value of tins, as opposed to tin, was only about £1 per ton.

On 10 July 1918, the *Times* carried an announcement from the National Salvage Council that fruit stones and nut shells were "needed at once by the Government for a special war purpose."[42] This cryptic request evidently failed to yield the desired result, for two weeks later another article appeared, which explained that they could be made into charcoal filters to protect British soldiers from chemical weapons. Anyone who wished to contribute could request a label from the government ensuring free shipping.[43] Two months later the organization called on local authorities to screen coal cinders from the refuse that they collected. If the heating value of these cinders were recovered, officials claimed,

[38] S. J. Langford to J. S. Oxley, 10 July 1916, NA, HLG 102/93.

[39] "War Uses of Old Iron: Munitions from Park Railings," *Times* (London), 15 Mar. 1918, 5.

[40] "National Salvage Campaign," *Times* (London), 2 Apr. 1918, 3.

[41] "£180,000 a Year from Old Tins," *Times* (London), 16 Apr. 1918, 3.

[42] "A Special War Purpose," *Times* (London), 10 July 1918, 3.

[43] "Fruit Stones and Nut Shells," *Times* (London), 26 July 1918, 8.

two million tons of coal would be conserved each year. This would not only save money but would also release many miners for military service or other war work. To promote this campaign, the council worked with the American labor organizer Samuel Gompers, who had recently arrived in Britain in an effort to boost the output and efficiency of British coal mines.[44] Two weeks before the Armistice, the National Salvage Council launched a campaign to collect old tires and other items made of rubber, which, the *Times* reported, were "urgently required."[45] Although salvage in Britain developed quite late in the course of the Great War and never approached the magnitude of that undertaken by the German and Austro-Hungarian empires, the experience provided lessons that would prove highly influential when the country returned to war a generation later.

SALVAGE BETWEEN THE WARS

Although some people in Britain had hoped that the salvage habit would remain strong after the Great War and thereby help the country to recover from the economic hemorrhage that had accompanied the literal loss of blood during the conflict, these dreams failed to materialize. Low prices discouraged recycling, and the government soon abolished wartime salvage mandates.[46] A small number of communities continued to operate salvage programs for household waste in the years that followed, but they reached only a tiny fraction of the nation's population. More significant was the scrap trade. From itinerant rag collectors to major demolition firms, many people earned their living from the collection and sale of waste. Writing in the early 1930s, H. G. Wells reported that one expert had informed him that "the Waste Trade, considered altogether, is the fifth greatest industry in England. It salvages everything from old iron, rusted girders, scrapped machinery and brick rubble, to bottles, bones, rags, worn-out tyres."[47]

During the Great Depression, demand for many commodities, both new and secondhand, virtually dried up. In May 1932, the cleansing department in Blackburn, Lancashire, possessed more than a hundred tons

[44] "Mr. Gompers on Coal and Victory," *Times* (London), 3 Sept. 1918, 3.
[45] "Waste Rubber," *Times* (London), 28 Oct. 1918, 2; "Scrap Rubber Salvage," *Times* (London), 4 Nov. 1918, 5.
[46] "Rubber Salvage Ended," *Times* (London), 2 Jan. 1919, 2.
[47] H. G. Wells, *The Work, Wealth and Happiness of Mankind* (London: William Heinemann, 1932), 215.

of pressed and baled cans, but it could not find a buyer at any price.[48] A number of communities that had collected household wastes ultimately decided to consign them to incinerators or landfills. Although much of the world remained mired in economic misery throughout the 1930s, by the middle of the decade two nations, Nazi Germany and Imperial Japan, appeared to be regaining their footing, prompted by enormous increases in military spending. To feed their weapons factories, Germany and Japan imported large quantities of scrap iron and steel from their future enemies, the United States and Great Britain. With both of these nations still in the grip of the Depression, their merchants eagerly welcomed this trade.

As German and Japanese rhetoric and deeds became increasingly bellicose, some in the democracies questioned the wisdom of supplying them with materials that they needed to make more weapons. Writing in 1937, the prescient British economist Alfred Plummer warned that the world seemed to be returning to the path that had led to the deaths of some nine million people just two decades earlier. "There has come, in the name of 'Defence,' a tremendous outburst of warlike preparations, which recall the situation in the years before August 1914, when all the major States of Europe were drawn into the vicious vortex of competitive arming, each protesting that it was none of their seeking and that they desired nothing but peace." Plummer feared that

> in the next war, if it ever comes, masses of men and materials will be swept to destruction on a scale unparalleled even in the last great war. Nearly all raw materials will become war materials. Whole nations will be conscripted and mobilised. Cities will be smashed like eggshells by fast bombing aeroplanes which, swooping out of the sky in numbers and at speeds never known before, will probably make the lives of non-combatants even more precarious than those of the combatants. Everywhere – at the front, behind the lines, "at home" – many thousands, perhaps millions, of men, women and children, of all ages, will be killed or maimed.[49]

Beginning in February 1937, the British government required merchants to obtain a license each time they wished to export iron or steel scrap.[50]

[48] M. Forcey to L. C. Hansen, 11 May 1932, and J. A. Webb to M. Forcey, 12 May 1932, both in Transport and General Workers' Union Collection, Modern Records Centre, Warwick University, MSS. 126/TG/RES/GW/74/1.

[49] Alfred Plummer, *Raw Materials or War Materials?* (London: Victor Gollancz, 1937), 143.

[50] "Export of Scrap Iron," *Times* (London), 27 Feb. 1937, 18. Similar bans had occurred during times of war or international crisis since the reign of Henry VIII. See Woodward, "Swords into Ploughshares," 190.

The intent of the new rules was not to hinder German rearmament but to ensure that sufficient raw materials would remain in the United Kingdom for its own weapons-production drive. Confident that British rearmament would soon lead to a sharp rise in steel prices, British merchants imported large amounts of scrap from the United States. These speculative purchases flooded the market, and the investors struggled to find buyers. In response, the British government allowed unregulated exports to resume in 1938. As one British metals dealer observed, merchants then "unloaded at a loss. They have been shipping, at a lot less than they paid the Americans, huge quantities to the Germans, for them to throw at us in the next war, I suppose."[51] British advocates of appeasement may have hoped that the resumption of scrap exports to Germany would lead Hitler to moderate his actions. If so, they were sorely mistaken.

In September 1938, just six months after he annexed his native Austria to Germany, Hitler again sought to redraw the map. His target this time was the Sudetenland, a region in Czechoslovakia that contained large numbers of German speakers. Desperate to avert war, the British prime minister, Neville Chamberlain, held a series of meetings with Hitler in Germany. Despite the fact that the Czechs were not even invited to the negotiations, Chamberlain and the French premier, Édouard Daladier, signed an agreement that gave Hitler the territory he demanded, which included most of Czechoslovakia's border defenses, airfields, and factories. Less than two months later, Nazi thugs went on a rampage across the newly enlarged Reich, killing hundreds of Jews, burning synagogues, and looting Jewish-owned businesses in what soon was dubbed *Kristallnacht* (the "Night of Broken Glass"). Instead of punishing these actions, the democracies reacted almost as if nothing had happened.

Although it is easy to condemn Chamberlain for his attempts to appease Hitler through one-sided concessions, he did so in an effort to prevent a war, which, if it came, would likely kill millions. Having lived through the tragedy of what many people then called "the World War," Chamberlain was passionately committed to seeking nonviolent solutions to international disputes. During the early 1930s, Chamberlain had fought to limit military spending, but by the time he became prime minister in May 1937 he strongly supported rearmament.[52] One year

[51] "'Outshots' . . . by the Rag Man," *Waste Trade World and the Iron and Steel Scrap Review* (hereafter *WTW*), 14 Jan. 1939, 1.

[52] Richard Overy, *The Twilight Years: The Paradox of Britain between the Wars* (New York: Viking Penguin, 2009), 343–4.

later, he oversaw the creation of an organization that would aid the injured and homeless in the event of an enemy attack on the cities and towns of Great Britain. In May 1938, the home secretary, Sir Samuel Hoare, approached Stella Isaacs, Marchioness of Reading, with an invitation to lead the new group, Women's Voluntary Services for Air Raid Precautions (the last three words were soon changed to "Civil Defence"). The wife of a former viceroy of India, Lady Reading had earned widespread admiration for her efforts to provide aid to unemployed people during the Depression. Hoare's message to her was blunt:

> We are faced with a danger such as we have never had to face before – the risk of attack by air which would be directed against the civil population of the country. . . . In times like these I feel that it is the duty of every citizen to help the nation to prepare against a danger which, if it comes, will affect all alike. The burden which would fall on women in the event of an attack would of necessity be heavy, and if it is to be met effectively it is essential that they should be organised and trained beforehand.[53]

Lady Reading agreed enthusiastically. Although WVS, as the organization quickly became known, had the appearance of being independent from the government, officials provided it with office space and an operating grant of about £15,000 a year – a figure which some civil servants initially questioned as extravagant.[54] They quickly relented, however. Taking note of "the very exceptional circumstances with which this organisation is to deal," the Treasury even agreed to pay retroactive wages to five of its staff members for the work they had done in the month before the organization officially began its existence. On 16 May 1938, WVS moved into a suite of offices in Queen Anne's Chambers, a building on Tothill Street in Westminster that was situated roughly halfway between Buckingham Palace and the Houses of Parliament.[55]

Just six months after WVS began, more than two hundred women worked in its London headquarters, and the group had tens of thousands

[53] Samuel Hoare to Lady Reading, 20 May 1938 (copy), NA, T 162/855. Hoare's comment overlooked the fact that German planes, zeppelins, and ships had killed scores of people in London, Hartlepool, and other British communities during the First World War. On Lady Reading, see Windlesham, "Isaacs, Stella, Marchioness of Reading and Baroness Swanborough (1894–1971)," in *Oxford Dictionary of National Biography* (Oxford: Oxford University Press, 2004).

[54] Wilfrid Eady to James Rae, 3 May 1938, NA, T 162/855; F. W. Holcombe [?], memorandum, 12 May 1938, NA, T 162/855.

[55] James Rae to Home Office, 30 June 1938 (copy), NA, T 162/855.

of volunteers across Britain.[56] Worried about its rapidly growing budget, one official complained that "the voluntary organisation for the recruitment of women volunteers which was to be assisted only with a contribution to cover rent, clerical assistance, travelling and office expenses, seems to have expanded into a Department of State with Lady Reading as the (unpaid) ministerial, and Miss Smieton as the permanent Head."[57] Chamberlain and his top officials did more than defend WVS against such criticism – they poured more resources into it. The reason was clear: war seemed highly probable, and the country was ill prepared for it. As WVS grew in size, so too did its responsibilities. In addition to the group's primary focus on civil defense, it would soon play a major role in the evacuation of children from cities and in the search for raw materials that could be recycled into weapons.[58]

During the 1930s, only a handful of British cities operated recycling programs. Then as now, the market for reclaimed materials was highly volatile, and many civic leaders doubted that the proceeds from the sale of recyclables would be sufficient to finance the costs of collecting them. Of the ten million tons of refuse that the British people discarded each year, all but a tiny fraction was burned, buried, or dumped into the sea. As late as January 1939, demand for waste paper was so weak that "hundreds and hundreds of tons of stock had to be thrown away."[59] Most people in Britain believed that recycling was simply not worth the trouble.

The situation was far different in the countries that would soon become known as the Axis Powers, which viewed recycling as vital to their economic and military strength. By 1937, the residents of all large German cities were required to salvage waste paper and animal bones, and members of the Hitler Youth were scouring the countryside for litter

[56] Lady Reading to Wilfrid Eady, 17 Nov. 1938 (copy), NA, T 162/855.

[57] F. P. Robinson to H. D. Hancock, 7 Dec. 1939, NA, T 162/855. Mary Smieton, a civil servant with the Home Office, served as the general secretary of WVS from 1938 to 1940.

[58] Virginia Graham, *The Story of WVS* (London: HMSO, 1959); Women's Voluntary Service for Civil Defence, *Report on 25 Years Work, 1938–1963* (London: HMSO, 1963); James Hinton, *Women, Social Leadership, and the Second World War: Continuities of Class* (Oxford: Oxford University Press, 2002). During its early years WVS (not *the* WVS) stood for "Women's Voluntary Services," but the group changed its name to the singular form after the war. By permission of its patron, Queen Elizabeth II, in 1966 it became Women's Royal Voluntary Service. In 2013 the group, which now focuses on helping elders, changed its name again, to Royal Voluntary Service. See www.royalvoluntaryservice.org.uk/about-us/our-history, accessed 23 March 2015.

[59] "Waste Paper Merchants' Annual Banquet in London," *WTW*, 14 Jan. 1939, 10–11.

that could be recycled.[60] The following year, in his foreword to a book published in Berlin entitled *Verwertung des Wertlosen* (The Exploitation of Waste Materials), Hermann Göring praised German chemists and engineers for the pioneering work that they had done to find uses for wastes. Interestingly, despite continued wartime paper restrictions, a British publisher issued an English translation of this work in 1944 with the title *Science and Salvage*.[61]

Japan, known in the 1930s as the "Land of No Waste," recycled corks, hats, and rubber – and imported large quantities of scrap metal from the United States.[62] In fascist Italy, rich and poor women alike helped Mussolini rearm by giving their gold wedding rings to the state, an act that the historian Victoria de Grazia argues "sparked a nationwide scrap-metal drive and gave a huge impetus to enlisting ordinary women in fascist institutions."[63] Italy and Germany also melted down tons of iron railings as they rearmed.[64]

The main reason that Germany, Japan, and Italy made recycling a matter of national policy in the 1930s was because they recognized their heavy dependence on other countries for many strategically important natural resources. Not surprising, many in the English-speaking world viewed this state of affairs as reassuring. *Britain To-Day*, a magazine produced by the British Ministry of Information, boasted in March 1939 that the United States and "members of the British Commonwealth of Nations" together controlled more than three quarters of the important minerals on the planet. The "blind chance of geology," asserted its author, "has turned in favour of those who stand for international cooperation, honourable dealing, and a respect for promises solemnly made by one State to another. The control of raw materials thus means, ultimately, that the law of nations is in safe keeping."[65]

60 "Germany Wastes Nothing," *Public Cleansing*, Dec. 1937, 146.

61 Claus Ungewitter, ed., *Verwertung des Wertlosen* (Berlin: Wilhelm Limpert, 1938); Claus Ungewitter, ed., *Science and Salvage*, trans. L. A. Ferney and G. Haim (London: Crosby Lockwood and Son, 1944).

62 A. L. Thomson, "Public Cleansing in National and in a War-Time Economy," *Public Cleansing*, May 1939, 365.

63 Victoria de Grazia, *How Fascism Ruled Women: Italy, 1922–1945* (Berkeley: University of California Press, 1992), 78.

64 "Germany's Economic Difficulties," *Times* (London), 19 Dec. 1938, 12.

65 "The Distribution of Raw Materials," *Britain To-Day*, 31 Mar. 1939, 6, 7. On British efforts to sway public opinion in the United States, see Nicholas John Cull, *Selling War: The British Propaganda Campaign against American "Neutrality" in World War II* (New York: Oxford University Press, 1995).

Yet G. A. Roush, a prominent American minerals expert, drew the opposite conclusion. In his view, the uneven distribution of raw materials might provoke aggressive behavior on the international stage. "The totalitarian states," he wrote in a 1939 book, "have been particularly concerned over their low degree of self-sufficiency and have been making strenuous efforts to remedy their shortcomings, which are much more pronounced than those of the democratic states. In fact, their marked lack of self-sufficiency has been one of the leading factors in bringing these states to a totalitarian form of government."[66]

On 15 March 1939, less than six months after Neville Chamberlain declared that his negotiations with Hitler had resolved "the Czechoslovakian problem," German troops marched into Prague. Heightened international tensions led many to fear that Britain might once again face the sort of material shortages that it had experienced during the Great War. As the leading journal of the British scrap trade observed, "Commodity Markets this week have all been under the dominating influence of the swift march of political events. . . . Rearmament must be intensified, and heavier requisitions for iron and steel are to be anticipated."[67] Two weeks later the same periodical reported that "users of all descriptions of iron and steel scrap complain of a shortage of supplies, the position in regard to heavy steel scrap being particularly acute."[68]

Many experts believed that Britain needed to work quickly to ensure that it would have the vital raw materials needed to fight a major war. Speaking to a national gathering of local officials in April 1939, Alexander L. Thomson, one of the leading waste experts in the United Kingdom, suggested that the British government should establish an organization similar to the National Salvage Council that had worked during "the last war" to ensure that wastes were reused.[69] Thomson's call brought no immediate action, but many shared his view that if war occurred, it would require careful management of raw materials from the outset. On 1 July 1939, two months before Germany invaded Poland, Conservative MP Leo Amery warned Chamberlain that "we are today in a much weaker position for blockading Germany than we were in the last war, while Germany is in a much stronger position for blockading

[66] G. A. Roush, *Strategic Mineral Supplies* (New York: McGraw-Hill, 1939), viii. Roush's words notwithstanding, the Soviet Union was of course no democracy.

[67] "Iron and Steel Scrap Markets," *WTW*, 25 Mar. 1939, 13.

[68] "Scrap Shortage," *WTW*, 8 Apr. 1939, 7.

[69] A. L. Thomson, "Public Cleansing in National and in a War-Time Economy," *Public Cleansing*, May 1939, 365.

us – and in 1917 she was not far from succeeding." Amery further observed, "So far we have accumulated certain limited stores of food against the opening weeks of confusion. . . . We have also, I know, accumulated certain reserves for the Fighting Services. But the risk of our industries, munitions and other, being brought to a standstill by lack of raw materials is no less serious and no less likely to be continuous. The Germans have made great preparations both by way of storage and by way of developing 'Ersatz' production. We have done nothing." Amery suggested that the situation was "so urgent that we ought to take immediate measures to see that every ship now afloat should at once fill up to fullest capacity with any kind of imported essential. . . . Whatever sums of money may be involved – and I dare say even a hundred million pounds would ensure our safety – it is only a small thing compared with our total armament expenditure."[70]

Attached to Amery's letter was a ten-page memorandum written by Arthur Salter, who had overseen British shipping during the Great War. Salter began by noting that in 1918 Britain had been "brought near starvation and the loss of the war through a reduction of imports and the absence of sufficient reserves of food and raw materials." He predicted that "our situation in another war would be much worse than in the last," for although the country had grown more dependent on imports, little chance existed that Britain would be able to obtain them in sufficient quantities in wartime. To prepare for the growing likelihood of war, Salter's advice was clear: "Buy and ship at once every possible ton of imported raw materials which can be dumped without deterioration or stored without difficulty, especially timber and ores and metals."[71]

Chamberlain accepted the need for action. Less than two weeks after he received the letters from Amery and Salter, he chaired a meeting of the Committee of Imperial Defence at which his government decided to import a million tons of iron ore, half of it from Sweden and half from French North Africa.[72] The prime minister also established a new government department, the Ministry of Supply, which formally began operating on 1 August 1939, one month to the day before Germany invaded Poland.[73] Chamberlain granted it sweeping powers over every raw

[70] Leo Amery to Neville Chamberlain, 1 July 1939, NA, PREM 1/375.

[71] Arthur Salter, "Reserves of Food and Raw Materials," 29 June 1939, NA, PREM 1/375.

[72] Ministry of Supply, Memorandum No. 35/39, 3 Sept. 1939, NA, POWE 5/64.

[73] John D. Cantwell, *The Second World War: A Guide to Documents in the Public Record Office*, 3d. ed. (Kew: Public Record Office, 1998), 177. For a detailed explanation of the structure of the Ministry of Supply, see James S. Earley and William S. B. Lacy, "British

TABLE 1. *Ministry of Supply*

Minister
Leslie Burgin (14 July 1939–12 May 1940)
Herbert Morrison (12 May 1940–3 Oct. 1940)
Sir Andrew Duncan (3 Oct. 1940–29 June 1941)
William Maxwell Aitken, first Baron Beaverbrook (29 June 1941–4 Feb. 1942)
Sir Andrew Duncan (4 Feb. 1942–26 July 1945)
John Wilmot (3 Aug. 1945–7 Oct. 1947)
Parliamentary Secretary
J. Llewellin, MP (14 July 1939–15 May 1940)
Harold Macmillan, MP (15 May 1940–4 Feb. 1942)
Wyndham Portal, Baron Portal (4 Sept. 1940–4 Mar. 1942)
Ralph Assheton, MP (4 Feb. 1942–30 Dec. 1942)
Charles Peat, MP (4 Mar. 1942–22 Mar. 1945)
Duncan Sandys, MP (30 Dec. 1942–21 Nov. 1944)
John Wilmot, MP (21. Nov. 1944–23 May 1945)
James de Rothschild, MP (22 Mar. 1945–23 May 1945)
Robert Villiers Grimston, MP (28 May 1945–27 July 1945)

Sources: John D. Cantwell, *The Second World War: A Guide to Documents in the Public Record Office*, 3d. ed. (Kew: Public Record Office, 1998), 177; *Times*, various dates; Ministry of Supply circulars held at local archives.

material with industrial or strategic significance as well as the authority to oversee the manufacture of munitions and other military supplies.[74] As the future prime minister, Harold Macmillan, would note less than a year later, "Both in its importance and the vastness of the field concerned, the Ministry of Supply is on a scale greater than that of any other department."[75]

Wartime Control of Prices," *Law and Contemporary Problems* 9, no. 1 (1942): 160–72, esp. 161–3.

[74] Some materials were under the purview of other government agencies. The Ministry of Aircraft Production controlled aluminum, for instance, and the Ministry of Food controlled foodstuffs. The Ministry of Supply also oversaw much of the nation's war production, although these responsibilities were shared by the Admiralty, the Ministry of Aviation, and (from 1942) the Ministry of Production. See Joel Hurstfield, *The Control of Raw Materials* (London: HMSO, 1953).

[75] Harold Macmillan to Herbert Morrison, 22 July 1940 (copy), Harold Macmillan Papers, Department of Special Collections and Western Manuscripts, Bodleian Libraries, University of Oxford, MS. Macmillan dep. c.267, fol. 189. Extracts from the Harold Macmillan Archive are reproduced with the kind permission of the Trustees of the HM Book Trust.

This proactive approach toward strategic resources was a sharp contrast to the laissez-faire ideology that had prevailed during the early years of the First World War. Embracing a policy of "business as usual," British government officials in 1914 had resisted the introduction of price caps, rationing, and compulsory recycling. Only toward the end of that war, as Britain faced significant shortages of many raw materials, did the government call on people to recycle for the war effort. By the 1930s, however, many believed that victory in any future war would require extensive intervention in the economy, including control over the use and disposal of vital materials.[76]

Yet Germany's demand for raw materials continued to be too strong for many foreign merchants to resist. In 1938 alone, American and British companies exported over half a million tons of scrap iron and steel to the Third Reich.[77] Many people in Britain found this deeply troubling, including a mother of three sons who asserted that the United Kingdom should retain every ounce of scrap metal it possessed. "If we have not room for it here," she declared a month after Chamberlain returned from his meetings with Hitler, it should be "sent to the bottom of the sea rather than to aid a foreign country to arm against us."[78] Robert Morrison, a Labour politician who represented the predominantly working-class constituency of Tottenham North in the House of Commons, expressed similar concerns about Britain's role in Germany's rearmament. "In the months preceding the war," he recalled in 1941, "when the Chamberlain Government, of accursed memory, was following a double policy of rearmament and appeasement, and making a mess of both, the Minister for the Co-ordination of Defence issued a stirring call to Local Authorities to collect scrap metal for armaments." Eager to do their bit, the people of Tottenham gathered over 600 tons of metal. Curious to see where it would be sent, Morrison and other residents of the borough accompanied the collection to a nearby wharf, where they were appalled to see it put on a ship bound for Germany. "Patriotic scrap merchants had sold the munitions material we had collected from door to door, to the enemy against whom the nation was rearming. When the department was

[76] Arthur Marwick, *Britain in the Century of Total War: War, Peace and Social Change, 1900–1967* (Boston: Little, Brown, 1968), 242, 268–70; Ina Zweiniger-Bargielowska, *Austerity in Britain: Rationing, Controls, and Consumption, 1939–1955* (Oxford: Oxford University Press, 2000).

[77] "The Scrap Position in Britain and Germany," *WTW*, 16 Dec. 1939, 10.

[78] "Germany Buys More English Scrap Metal," *Sunday Express*, 23 Oct. 1938, 8, clipping in NA, SUPP 3/49.

informed, the complacent reply was that it was unfortunate, but could not be avoided."[79] In the summer of 1939, German buyers negotiated a number of such deals, and the press ominously reported that "the insistence placed by Germany on early delivery adds special significance to the purchases, in view of the current international situation."[80]

On 23 August 1939, Nazis and Communists temporarily set aside their ideological differences for an alliance of expedience, the German-Soviet Nonaggression Pact. Despite its peaceful-sounding title, many people recognized this treaty for what it was: a prelude to war. Signed by Joachim von Ribbentrop and Vyacheslav Molotov, the respective foreign ministers of Hitler and Stalin, this treaty included a secret protocol in which these two dictators agreed to jointly occupy Poland and divide it between them. That very day, the British government instituted strict export controls that put an immediate halt to any further shipments of scrap metals to Germany.[81] In addition to regulating exports, the Ministry of Supply imposed strict controls over the sale and use of many materials within the domestic economy. Anxious to avert rampant inflation of the kind that had taken place during the previous war, the ministry also established stringent price caps. When subsequent developments prompted it to increase the controlled price of a particular commodity, it simultaneously imposed a tax on inventories to prevent windfall profits from occurring and to discourage speculative purchases that might lead to hoarding.[82]

In late August, as war seemed increasingly likely, a sixty-year-old British civil servant named J. C. Dawes made a novel suggestion. "In the event of a national emergency," he informed one of his superiors, "the question of salvage and utilisation of waste materials" could play a vital role in supplying Britain with the materials needed to fight against Nazi Germany. Dawes spoke from experience, for he had served as Britain's deputy director general of salvage during the Great War. In that capacity he had overseen the recycling of metal, paper, and food waste in communities across Britain. Following the war, Dawes had joined the Ministry of Health, where he was responsible for supervising the nation's municipal refuse collection services. As the threat of war loomed, Dawes

[79] R. C. Morrison, "Paper, Pigs, Poultry – and Any Old Iron," *Tribune* (London), 8 Aug. 1941, 7.
[80] "Government and Essential Supplies," *WTW*, 26 Aug. 1939, 2.
[81] "Transactions in Copper," *WTW*, 26 Aug. 1939, 11.
[82] E. Leslie Burgin, "Eighth Monthly Report of the Minister of Supply covering the Month of March 1940," 20 Apr. 1940, 8, NA, CAB 68/6.

viewed recycling as essential to national security. "Having regard to the difficulties of obtaining supplies in 1917–18," he explained, "I have encouraged the erection of refuse disposal plants . . . designed to make salvage in an emergency possible. On the word 'GO' we could take from this quantity of refuse any material required for war purposes."[83]

[83] J. C. Dawes to R. G. Hetherington, 29 Aug. 1939, NA, HLG 51/556. On Dawes's long career, see "The Salvage of Refuse: Materials Needed by Industry," *Times* (London), 1 Dec. 1939, 10; J. F. M. Clark, "Dawes, Jesse Cooper (1878–1955)," in *Oxford Dictionary of National Biography* (Oxford: Oxford University Press, 2004).

2

Persuasion and Its Limits

> The problem to-day is to beat our ploughshares into swords, at the same time providing additional ploughshares, so that industry can supply civilian needs and carry on export trade.
>
> –*The Times*, 1939[1]

On 1 September 1939, German forces invaded Poland, beginning the Second World War in Europe.[2] When Hitler ignored British and French demands that he withdraw, both countries declared war on Germany. In the midst of Prime Minister Neville Chamberlain's live radio address on 3 September to announce that a state of war existed between the United Kingdom and the Third Reich, air raid sirens sounded in London. The sirens proved to be a false alarm, but just hours later, a German U-boat torpedoed a British passenger liner, the SS *Athenia,* as it steamed through the North Atlantic toward Montreal. Most on board were rescued before the ship went down, but more than a hundred passengers and crew members lost their lives. Six weeks later another U-boat sank the British battleship HMS *Royal Oak* at its base at Scapa Flow in Scotland, killing 833 sailors.[3]

In contrast to this bloodshed at sea, late 1939 and early 1940 passed surprisingly quietly on the land and in the skies of Western Europe. When the expected attacks on their city failed to materialize, many Londoners

[1] "Ploughshares into Swords," *Times* (London), 7 Nov. 1939, 5.

[2] The Czechs might disagree. The starting date of the war in the Pacific is even more contested.

[3] Peter Padfield, *War beneath the Sea: Submarine Conflict during World War II* (New York: John Wiley, 1998).

who had sent their children to places of safety brought them home; others criticized the government for paying air raid wardens who had little to do. To prepare Britons for the hardship and sacrifice to come and to convince them that wartime infringements on their freedom were necessary, government officials needed to show that they were taking every possible step to organize the country for the difficult challenges of fighting Nazi Germany. At the same time, they had little confidence that civilians would remain calm if Hitler subjected them to bombs and possibly poison gas. Voluntary recycling, they believed, would boost morale by making civilians feel personally engaged in the war. Almost as an afterthought, they realized that salvage might also provide useful materials for the war effort. There was certainly room for improvement. At the start of the war the United Kingdom was consuming well over three million tons of paper annually, yet its towns and cities were recycling only 50,000 tons of waste paper each year – less than 2 percent of the total.[4] Private scrap merchants collected considerably more than did local authorities. According to a government estimate, the British scrap trade recycled 700,000 tons of waste paper annually in the late 1930s.[5]

Britain's lackadaisical approach to recycling during the opening months of the war stood in sharp contrast to the forceful way the government regulated consumption. Emergency wartime rules affected the export, sale, and use of practically every commodity, and they capped prices at levels considerably lower than those on the international market. The goal was to hold down the cost of rearmament and prevent inflation. As an official in the United States War Department observed in early October 1939, "The economic regimentation of Britain has reached a point, after only four weeks of war, which took two years to bring about in the 1914–1918 conflict. An immediate State control of the essential factors of the nation's economy was initiated during the first fortnight of September and there has been no attempt this time to maintain 'business as usual' in conjunction with the current wartime economic requirements."[6] Parliament gave the government almost unlimited authority to create agencies and enact

[4] "Waste from Refuse," *Waste Trade World and the Iron and Steel Scrap Review* (hereafter *WTW*),13 Jan. 1940, 9; C. R. Moss, "Reclamation of Waste Materials from Refuse in War Time," *Public Cleansing*, Jan. 1940, 133, 136; Waste Paper Recovery Association, *Annual Report, 1955* (London: n.p., 1956).

[5] See Inter-Departmental Committee on Salvage, Minutes, 13 May 1947, 5, NA, BT 258/404.

[6] War Department, "The First Month of Britain's War Economy," 7 Oct. 1939, 1, Papers of Edward R. Stettinius Jr., 1924, 1949–50, Accession #2723-z, Special Collections, University of Virginia Library, Charlottesville, Va. (hereafter Stettinius Papers), Box 75.

regulations without having to codify them in legislation. Visiting Britain during the first month of the war, the American economist Willard E. Atkins was struck by hearing the remark "that England, overnight, has become a totalitarian state. . . . Nevertheless it was realized that to defeat Germany it is necessary to do much of what Germany has already done."[7] Wartime Britain was hardly totalitarian, but the powers the government assumed were nonetheless unprecedented and extremely far-reaching.

Shortly after Britain went to war, the "Rag Man," author of a weekly column in the magazine *Waste Trade World and the Iron and Steel Scrap Review*, weighed in on the price controls that the government had recently imposed. Although he explained that he did not welcome bureaucrats telling the industry how to operate, he maintained that controls were necessary to prevent owners of strategically vital commodities from holding them ransom in the expectation that prices would rise, as had happened during the war of 1914–18. Most members of the trade grudgingly accepted this assessment.[8] Two months later, however, the Rag Man complained that the prices the government had established were so low as to be unfair to the scrap trade. Merchants, he argued, had so little incentive to sell that the government had been forced to abandon its reliance on market forces to create equilibrium between supply and demand. The Rag Man denounced the government's reliance on local officials and volunteers to gather salvage and asserted that "the complications of the Waste Reclamation Industry are too manifold to be grasped in a few days by enthusiastic amateurs."[9] This critique was merely an early salvo in a debate that would continue throughout the rest of the war, and indeed for decades after it, about the role of the state in the economy.

Faced with the burden of importing unprecedented quantities of strategic materials at high prices, Britain's trade deficit had increased sharply during the last year of peace. With the outbreak of war, reliance on imports posed dangers of an entirely new character. The people of Britain, well aware of the devastating effects of the British naval blockade against Germany during the previous war and newly cognizant of the threats that

[7] "Control of War Materials," *WTW*, 9 Sept. 1939, 2; J. C. Carr, memorandum, 25 Nov. 1939, NA, POWE 5/104; Willard E. Atkins, et al., "Confidential Report on Great Britain's Industrial and Financial Mobilization Plan, Submitted to the War Resources Board by the Brookings Institution, October 13, 1939," 31, 105–6, Stettinius Papers, Box 76.

[8] "'Outshots' . . . by the Rag Man," *WTW*, 16 Sept. 1939, 2

[9] "'Outshots'. . . by the Rag Man," *WTW*, 25 Nov. 1939, 2.

Hitler's U-boats posed to Allied shipping, were "being recalled to habits of prudence and economy. . . . The lives of our seamen must not be needlessly endangered, and shipping must be saved for essential war needs and absolute necessities."[10] At the same time, foreign trade remained essential. Although production for export consumed precious materials, labor, and manufacturing capacity, it also generated the foreign currency that Britain needed to pay for vital raw materials that it could obtain only from outside the British Empire.[11]

At the start of the war, Britain's fleet of large oceangoing vessels, excluding tankers, stood at 14.3 million gross tons. Over the next sixteen months alone, German attacks destroyed a fifth of these ships. One of the worst single days for Allied shipping occurred in May 1941, when German submarines sank eight merchant vessels near the southeast coast of Greenland.[12] "In view of the shipping position," noted an internal Ministry of Supply memo that month, "particular attention has been paid to the problem of expanding the collection of home scrap."[13] The human toll was equally grim. One out of every three British merchant seamen was killed during the war – more than 30,000 in total. Despite frantic efforts to build new ships and to salvage vessels that had been sunk, at the end of 1940 Britain's cargo capacity was 1.4 million gross tons smaller than it had been at the start of the war. As a result, the British had to rely increasingly on foreign vessels to bring the food and raw materials they needed to survive.[14]

Finding willing firms proved difficult, for Britain paid a lower rate than did some other countries despite the considerable risks involved in sailing to its shores. Not surprisingly, as an official with the Ministry of Shipping observed in March 1940, "Operators of neutral ships not only fail to show an interest in our priority cargo, but are inclined more than ever to avoid our ports." Even more vexing than neutral ships were foreign-owned ships registered in British ports. One such case was the SS *Gypsum*

[10] "A Great Role – Stopping the Leaks in National Waste," *Public Cleansing*, May 1940, 267.

[11] Leo M. Cherne, William J. Casey, and James L. Wick, "An Analysis of the Mobilization of Economic Resources for War in Germany and England," Research Institute of America, [1940], 56, Stettinius Papers, Box 87.

[12] Averell Harriman to Harry Hopkins, 21 May 1941 (copy), W. Averell Harriman Papers, Manuscript Division, Library of Congress, Washington, D.C. (hereafter Harriman Papers), Box 159, Folder 5.

[13] "Iron and Steel Scrap," 14 May 1941, NA, POWE 5/60.

[14] "The Imports Situation," War Cabinet, 29 Jan. 1941, W.P. (41) 17, Cherwell Papers, Nuffield College Library, Oxford, F.254/1/2.

Queen, owned by the Chicago businessman Otis Wack. When the Ministry of Shipping ordered him to carry a load of steel to the United Kingdom in 1940, he "rebelled violently." Wack's obstinacy led one British official in New York to denounce him as "a very tough customer indeed. . . . Mr. Wack, although he has enjoyed for many years the privileges that go with the British flag on his steamers, adopted a purely Chicago attitude that somebody else's war had no right to interfere with his business."[15] Wack eventually gave in to this pressure; on 11 September 1941, a German submarine, U-82, sank the *Gypsum Queen* off Greenland while it was en route to Liverpool with 5,500 tons of sulphur. Ten crew members lost their lives.[16]

The Exchange Requirements Committee, which operated within the Treasury, had the responsibility of estimating how much money government departments would need to expend on foreign supplies. Imports from the empire brought substantial cost savings, but they could not meet Britain's voracious demand for scrap, the primary overseas supplier of which was the United States.[17] In the fall of 1939 alone, Britain ordered half a million tons of steel scrap from U.S. suppliers.[18] Britain's imports of American scrap in the first six months of 1940 were three times greater than during the same period the previous year.[19] These imports depleted Britain's dwindling reserves of dollars and required a perilous journey through waters swarming with U-boats.

Just weeks after the war began, the National Federation of Scrap Iron and Steel Merchants, acting in conjunction with the Ministry of Supply, sent letters to the heads of industrial firms throughout Britain asking them to scour their premises for scrap iron and steel. Lest the public become alarmed, a correspondent in the *Times* declared, "This is not a panic measure designed to obtain supplies of scrap for steelworks that are unable to obtain sufficient to carry on. The raw material position is strong. There is no necessity, as is the case in Germany, to approach householders to hand over odd pieces of metalware. An

[15] I. M. Evans to Julian Foley, 29 Mar. 1940, NA, MT 59/849. In May 1941 the Ministry of Shipping and the Ministry of Transport merged to form the Ministry of War Transport.

[16] Jürgen Rohwer, *Axis Submarine Successes of World War Two: German, Italian, and Japanese Submarine Successes, 1939–1945* (Annapolis, Md.: Naval Institute Press, 1999), 65; Arnold Hague, *The Allied Convoy System, 1939–1945: Its Organization, Defence and Operation* (Annapolis: Naval Institute Press, 2000), 136.

[17] Exchange Requirements Committee, Minutes, 25 Apr. 1940, NA, T 161/1405.

[18] "Scrap for Britain," *WTW*, 23 Dec. 1939, 2.

[19] "Britain Buys U.S.A. Scrap," *WTW*, 31 Aug. 1940, 3.

acceleration of supplies of scrap from regular sources and from otherwise neglected dumps will reduce the risk of ocean transport and will save foreign exchange."[20] Leaders of the campaign claimed that by the end of 1939 this appeal had brought forth 80,000 tons of scrap metal from British businesses – enough to free up as many as a dozen ships that would have been needed to bring this material to Britain had it been imported.[21]

Instead of viewing such efforts as excessive, many Britons thought that their government should do much more to utilize neglected resources. As the *Manchester Guardian* observed in November 1939, "In every armament factory, every industrial works, in builders' yards, as well as in the households of this country there are enormous reserves of valuable raw materials. Germany has been combed for years past for anything useful, from church bells to iron railings, from kitchen refuse to toothpaste tubes."[22] A short time later the *Times* echoed this point and quoted from a letter that an unnamed workman had written to its editor: "Believe me, Sir, to be a loyal, true, and faithful British subject, but I mention one good point re that swine Hitler – he is no waster."[23]

In contrast to Nazi Germany, where recycling had become compulsory in large towns and cities well before the war began, British officials hoped at first that mandatory recycling would not be necessary. At the start of the war, the British press had pointed to German laws that made it a criminal offense to waste food or drink as emblematic of Nazi despotism. Less than a year later, however, the *Times* welcomed the British government's decision to make it a crime to waste food. It added that a similar regulation had existed in the United Kingdom during the 1914–18 war.[24] Despite this prohibition against wasting food, British officials refrained from extending a similar provision to the destruction of other materials until 1942.

[20] "National Scrap Campaign" *WTW*, 23 Sept. 1939, 2.

[21] "Crusade against Waste," *Times* (London), 10 Jan. 1940, 7; "Scrap Metal Campaign," *Times* (London), 30 Jan. 1940, 3.

[22] "Wealth from Waste: A Salvage Campaign," *Manchester Guardian*, 13 Nov. 1939, 9. See also Adam Tooze, *The Wages of Destruction: The Making and Breaking of the Nazi Economy* (New York: Viking, 2006), 214.

[23] "Great Salvage Drive," *Times* (London), 3 Feb. 1940, 3. On the role of this and other British newspapers during the Second World War, see Stephen Koss, *The Rise and Fall of the Political Press in Britain, Vol. 2: The Twentieth Century* (Chapel Hill: University of North Carolina Press, 1984), esp. 592–603.

[24] "Food Criminals," *Times* (London), 7 Aug. 1940, 5.

Rather than wait for officials in London to require them to recycle, some civic leaders had already taken the initiative. Only a month after the start of the war officials in Edinburgh began asking residents to refrain from throwing away paper and encouraging them to leave it in separate bundles for collection.[25] In November the town clerk of Newbury wrote to the Ministry of Health to ask whether the local authorities should collect tin cans and other recyclable materials for the war effort. An official replied that the matter was under discussion and indicated that the government was preparing to issue salvage instructions to local government authorities.[26]

Meanwhile, in the Third Reich, arms production had grown to such a high level that Germany faced shortages of iron and steel. To obtain it, authorities in the Reich intensified their efforts to reclaim civilian articles and melt them down for military use. Supplies of scrap metal were so scarce in Germany that Nazi officials collected iron railings, signs, lampposts, and iron decoration from buildings. As in Britain, some in Germany claimed that doing away with superfluous ornamentation would "improve the aesthetic appearance of . . . houses and towns."[27] Germany's top salvage official declared that not a single item could be overlooked. Every house was to have separate containers for paper, bones, rags, and metal, which would be collected by members of the Hitler Youth.[28] So seriously did Nazi authorities view scrap metal that they executed four Polish workers in the German-occupied city of Poznań for allegedly stealing metal from the scrapyard where they worked.[29]

In March 1940, Hermann Göring called on the German people to collect scrap metal in honor of Hitler's birthday. The *Times* observed that "the 'voluntary contribution' will presumably be a great turning out of pots and pans, for most of the 'visible reserves' of metal, such as street and park railings, disappeared some time before the war started."[30] When municipal officials in Edinburgh heard about Göring's efforts, they declared that the Nazi leader's "birthday should be marked in this country by a wholesale

[25] "Value in Waste Paper," *Times* (London), 2 Oct. 1939, 2.

[26] See S. Widdicombe to the Ministry of Health, 17 Nov. 1939; J. C. Dawes to I. F. Armer, 21 Nov. 1939; and J. Poyser to S. Widdicombe, 24 Nov. 1939, all in NA, HLG 51/556.

[27] "In Germany To-Day," *Times* (London), 2 Jan. 1940, 7.

[28] "Prevention of Waste in Germany," *Public Cleansing*, Apr. 1940, 259.

[29] "We Have Not Reached This!" *Public Cleansing*, May 1940, 298.

[30] "Birthday Presents for Hitler: Scrap Metal Campaign," *Times* (London), 16 Mar. 1940, 7.

response to the appeals for 'more salvage.' All the scrap collected would be hurled at, and not sent to, Hitler for his birthday!"[31]

The government of fascist Italy also looked to iron railings as a source of raw material. In April 1940, the *Times* reported that the Italian cabinet had approved a measure requiring owners of railings to sell them "to the State as scrap-iron for use in the making of armaments." The article noted, however, that the new law exempted railings of historic or artistic value as well as those connected with religious sites.[32]

Although some people in Britain hoped their government would emulate authoritarian countries and require citizens to recycle everything that could be used in the war effort, others argued that compulsion would infringe on civil liberties and might even lead the government to spy on what people put in their rubbish bins. Speaking to a group of municipal waste disposal experts in April 1940, the deputy controller of salvage, J. C. Dawes, declared that he did not think compulsion would be necessary to induce the public to recycle. He added, however, that the government would not hesitate to make salvage mandatory if a voluntary approach failed to yield sufficient results.[33]

Soon Britain would collect railings with an even greater fervor than its authoritarian enemies. Professor A. E. Richardson, a leader in the Georgian Group, a preservation society dedicated to saving the architectural heritage of the eighteenth and early nineteenth centuries, declared in July 1940 that most railings "erected after 1850 could wisely be spared for scrap." Echoing the arguments of those who sought to exercise discrimination in determining which historical documents should be pulped, Richardson asserted that selective preservation would not hamper the country's war effort, because "good iron-work" constituted only a tiny percentage of the total.[34] Although the members of the Georgian Group hoped to preserve what they considered most precious, this was a Faustian bargain, for it eroded their independence and made them complicit in the destruction that wartime salvage continued to cause. James Melvin, an official with the Iron and Steel Control agency who played a key part in the government's efforts to melt railings, welcomed the

[31] "Edinburgh's Campaign," *Public Cleansing*, Apr. 1940, 259.
[32] "Civil Mobilization in Italy," *Times* (London), 3 Apr. 1940, 8.
[33] "Salvage of Paper and Textiles: New Importance," *Manchester Guardian*, 13 Apr. 1940, 5.
[34] A. E. Richardson, letter to the editor, *Times* (London), 30 July 1940, 5. Several months later the secretary of the Society for the Protection of Ancient Buildings made a similar plea. See John E. M. MacGregor, letter to the editor, *Times* (London), 18 Nov. 1941, 5.

involvement of the Georgian Group for precisely this reason. Significantly, as early as the summer of 1940 Melvin suggested that military necessity might eventually require even the most significant railings to be melted down. Writing to the *Times*, he commented that "beautiful and worthy" railings "should at least be left until the very last in the present urgent search for raw materials. In reporting to us as many cases as possible of ugly and redundant ironwork, the [Georgian] Group can ensure that the opportunity for removal is not missed."[35]

THE SALVAGE DEPARTMENT

On 5 October 1939, barely a month after Britain declared war against Germany, Chamberlain announced that he was considering establishing a department devoted entirely to the reclamation of wastes.[36] Six days later, J. C. Dawes met with officials from a number of agencies, including the Admiralty, the War Office, the Air Ministry, and the Ministry of Food. All supported the creation of a salvage department in Whitehall. Dawes explained that during the First World War, "The National Salvage Council had had only a small staff at headquarters and had used the services of the leading Cleansing [i.e., municipal waste management] officials as honorary advisers all over the country. The same could be done again."[37] On 11 November 1939, exactly twenty-one years after the end of the First World War, the *Times* announced that the government had established a Salvage Department within the Ministry of Supply.[38] Despite Dawes's role in shaping the new department, he was not chosen to lead it. The top position went to Harold Judd, a partner in the accounting firm of Mann, Judd, Gordon, and Co. who had spent the 1914–18 war working in the contracts department of the Ministry of Munitions. Judd quickly asked Dawes to oversee municipal salvage collection, a position the latter would hold throughout the remainder of the war.[39]

[35] James Melvin, letter to the editor, *Times* (London), 31 July 1940, 5.

[36] "Parliament," *Times* (London), 6 Oct. 1939, 3.

[37] Minutes of the second meeting of the salvage committee, Ministry of Supply, 11 Oct. 1939, NA, HLG 51/556.

[38] "Using up Waste Products: Task of New Salvage Department," *Times* (London), 11 Nov. 1939, 8.

[39] "A Big Task: Where the Cleansing Service Comes in," *Public Cleansing*, Jan. 1940, 144. Harold Godfrey Judd (1878–1961) does not appear in the *Oxford Dictionary of National Biography*, but the National Portrait Gallery owns two photographic portraits of him. His obituary notes that he served as mayor of Hampstead from 1951 to 1953. See "Mr. Harold Judd," *Times* (London), 7 Jan. 1961, 10.

TABLE 2. *Salvage Departments*

Position	Holder
Salvage Department (Oct. 1939–Oct. 1941)	
Controller of Salvage	Harold Judd
Deputy Controller of Salvage	J. C. Dawes
Directorate of Salvage and Recovery (Oct. 1941– Aug. 1942)	
Director of Salvage and Recovery	G. B. Hutchings
Controller of Salvage	Harold Judd
Controller of Salvage (Local Authorities)	J. C. Dawes
Directorate of Salvage and Recovery (Sept. 1942–Feb. 1945)	
Principal Director of Salvage and Recovery	G. B. Hutchings
Director of Salvage and Recovery	Harold Judd
Director of Salvage and Recovery (Local Authorities)	J. C. Dawes
Directorate of Salvage and Recovery (Feb. 1945–Feb. 1946)	
Director of Salvage and Recovery	Harold Judd
Director of Salvage and Recovery (Local Authorities)	J. C. Dawes
Directorate of Salvage and Recovery (Feb. 1946–Mar. 1950)	
Director of Salvage and Recovery	J. C. Dawes (Feb. 1946–ca. Aug. 1948) N. S. W. Ashwanden (ca. Aug. 1948–Mar. 1950)

Sources: Times (London), various dates; Salvage Circulars held at local archives.

Judd promised that he would take a comprehensive approach to salvage, one that would include wastes from homes, factories, and the armed forces. Because industry was already recycling considerable quantities of its own wastes and because members of the military could be ordered to recycle, salvage officials directed most of their energies toward households. The central government clearly lacked the resources necessary to run a nationwide recycling program on its own, so the Salvage Department asked the local authorities that governed Britain (borough councils, county borough councils, urban district councils, and rural district councils) to encourage residents to separate useful materials from the rest of their rubbish. Officials in London hoped that the task of gathering recyclables could simply be added to the work of municipal rubbish collection. This was more easily said than done. Labor was tight, as many dustmen had enlisted in the armed services or taken jobs in war factories. Few towns or cities owned vehicles that would allow salvage to be kept separate from rubbish. Some communities eventually collected rubbish and salvage on alternate weeks; during special salvage drives, intervals of three weeks between rubbish collections were not unheard of.[40]

On 30 November 1939, the Soviet Union invaded Finland, beginning a bloody three-and-a-half month conflict known as the Winter War.[41] Finland was an important source of wood pulp for Britain's paper industry. Within days of this invasion, approximately fifteen hundred local authorities in England, Scotland, and Wales received a letter from the Ministry of Health introducing "Salvage Circular 1" from the newly constituted Salvage Department of the Ministry of Supply.[42] The reason the new circular came under this cover was that local authorities were accustomed to receiving directives from the Ministry of Health and were unlikely to overlook something that it issued. In contrast to later appeals that would offer explicit examples of the military significance of recycling, this circular asserted that the nation needed salvaged items for the manufacture of civilian articles such as glue, fertilizers, and animal feed. The document also ordered all but the smallest localities to begin submitting monthly reports about the types and amounts of materials they collected for recycling. Significantly, this directive from the Ministry of Supply made no mention of "rag-and-bone men," scavengers who had long earned a living from

[40] E. Leslie Burgin, "Fourth Monthly Report by the Minister of Supply Covering the Month of November," 18 Dec. 1939, NA, CAB 68/3/29; "Salvage Department," *WTW*, 23 Dec. 1939, 4; "Salvage Controller," *Times* (London), 20 Nov. 1939, 6; "Getting a Move on," *Public Cleansing*, Jan. 1940, 120; "Government Salvage," *WTW*, 25 Nov. 1939, 3.

[41] Norman Davies, *No Simple Victory: World War II in Europe, 1939–1945* (New York: Penguin, 2007), 79–81.

[42] Ministry of Health, Circular No. 1919 (30 Nov. 1939), NA, HLG 51/556.

the house-to-house collection of old or worn-out items that they could resell for a slight profit.[43]

The Ministry of Supply soon issued two additional circulars, one of which admitted that "the difficulty of obtaining wood pulp . . . is now giving rise to serious difficulties."[44] To address this problem the deputy controller of salvage, J. C. Dawes, called on towns to increase by a factor of five the amount of paper they collected for recycling, to reach an annual rate of 250,000 tons.[45] In contrast to Dawes's insistence that salvage could make a major contribution to the war effort, the minister of supply, Leslie Burgin, displayed only moderate interest in recycling. Although Burgin quickly instituted rationing and price controls, he refused to make salvage compulsory for either municipalities or individuals.[46]

To supplement the salvage work of municipal dustmen, the Salvage Department decided to ask Women's Voluntary Services (WVS) for help. This was a natural place to look, for in the month of September 1939 alone, more than 100,000 women had joined its branches throughout Britain. By the end of March 1940, WVS had recruited more than 600,000 volunteers.[47] That month, an article in its magazine, the *Bulletin*, observed, "The Ministry of Supply, having strengthened its faith in the W.V.S., is advising one borough after another to address themselves to the W.V.S. for help towards Salvage. We are now helping Paddington, Westminster, Hackney, Wembley, Romford, Dartford – all with satisfactory results." WVS members promoted salvage by handing out leaflets and speaking directly to other women about the importance of salvage. Reporting on this effort, the *Bulletin* explained, "That is our major work – getting into contact with the housewife and teaching her to be Salvage-minded."[48] During the first six months of 1940, WVS members visited half a million households in London to spread the word about the need to recycle.[49]

[43] Ministry of Supply Salvage Department, Salvage Circular 1 (Nov. 1939), NA, HLG 51/556. For a discussion of this circular, see George North to J. C. Carr, 23 May 1940, NA, HLG 102/92.

[44] Ministry of Supply Salvage Department, Salvage Circular 3 (Jan. 1940), Shakespeare Centre Library and Archive, Stratford-upon-Avon, BRR 55/14/31/3/1.

[45] C. R. Moss, "Reclamation of Waste Materials from Refuse in War Time," *Public Cleansing*, Jan. 1940, 133, 136.

[46] "Salvage and Supply," *Manchester Guardian*, 13 Apr. 1940, 8. There is no entry in the *Oxford Dictionary of National Biography* for Burgin (1887–1945), but his death did not go without notice; see "Dr. Leslie Burgin," *Times* (London), 17 Aug. 1945, 7.

[47] "Statistic Department," WVS *Bulletin*, Nov. 1939, 7; "Statistical Survey of W.V.S. Enrolments," May 1940, 2.

[48] "Salvage," WVS *Bulletin*, Mar. 1940, 8–9.

[49] "Salvage," WVS *Bulletin*, July 1940, 4.

Most towns and cities operated rubbish collection services to which recycling could be appended, but rural areas did not. People in the countryside generally burned their rubbish and buried what remained. This created a challenge for officials. How could they make it possible for people on farms and in villages to contribute to wartime salvage efforts? As an alternative to house-to-house collection, the Ministry of Supply asked rural authorities to create public collection points, called village dumps, to which people were supposed to bring recyclable materials.[50] For help in organizing and clearing these dumps, the Salvage Department approached Lady Gertrude Denman, president of the National Federation of Women's Institutes, a largely rural organization that been created in 1917 with the goal of boosting domestic food production. In a letter to Denman, Judd wrote that he was trying to "stimulat[e] the recovery of waste material and its conversion into productive material for war purposes." The Salvage Department, he explained, had already been in contact with the Boy Scouts and Women's Voluntary Services, and it now sought the cooperation of her federation.[51] The group's general secretary, Frances Farrer, quickly replied, promising to help collect salvage from villages throughout England and Wales.[52]

Despite the government's efforts to encourage recycling, most people continued to throw away their rubbish without giving it any thought.[53] Aware of this, waste management expert A. L. Thomson complained in February 1940 that "through neglect or wantonness, considerable quantities of useful material are carelessly discarded. In a time of stress like the present such action is traitorous."[54] That month, the ministers of supply and health chastised the lord mayor of the City of London and the mayors of most of London's metropolitan boroughs for failing to promote

[50] *Yorkshire Post*, 9 Mar. 1940; Note on the Collection of Scrap Iron from Farms, 4 July 1940, NA, MAF 58/226; Central Parish Councils Committee, *Salvage in Rural Areas* (London: National Council of Social Service, 1944).

[51] Harold Judd to Lady Denman, 6 Jan. 1940, Records of the National Federation of Women's Institutes, from Special Collections and Archives, London School of Economics, 5FWI/A/3/073; Tessa Stone, "Denman, Gertrude Mary, Lady Denman (1884–1954)," in *Oxford Dictionary of National Biography* (Oxford: Oxford University Press, 2004).

[52] Harold Judd to Frances Farrer, 16 Jan. 1940, Records of the National Federation of Women's Institutes, Special Collections and Archives, London School of Economics, 5FWI/A/3/073; Julie Summers, "Farrer, Dame Frances Margaret (1895–1977)," in *Oxford Dictionary of National Biography* (Oxford: Oxford University Press, 2004).

[53] "Salvage and Supply," *Manchester Guardian*, 13 Apr. 1940, 8.

[54] A. L. Thomson, "Exploring the Possibilities of Waste Materials," *Public Cleansing*, Feb. 1940, 184.

recycling: "In December, we regret to say, the value of materials salved from house refuse by the Metropolitan Borough Councils was only £1,670, and seven-eighths of this was salved in 5 out of the 29 Metropolitan Boroughs. In December also London salved only 140 tons of clean paper. Edinburgh with one-eighth of London's population saved 600 tons – more than 4 times as much."[55] About the same time, a resident of Bedford Square complained that although she had tried "not to throw into the dustbin anything that might be of material use," the officials in her borough thwarted her efforts. "My flat is choked with unsightly bundles and I cannot get anyone to take these away. . . . I am told that if I lived in Tottenham or Finsbury eager local authorities would collect and ask for more, but I live in Holborn, and Holborn cares for none of these things."[56]

In March 1940, an official with the Ministry of Supply told the War Cabinet that the country faced a looming shortage of many strategically important raw materials.[57] Britain's steel industry was operating with stocks sufficient to last less than two weeks.[58] The shortage of steel was so acute that despite the serious threat of air attack, the government suspended the manufacture of Anderson shelters. Made from curved sheets of corrugated steel that were bolted together, these shelters were designed to provide bomb protection in people's back gardens against everything but a direct hit. The timing of this suspension could not have been worse, for it occurred fewer than six months before Hitler launched massive bombing raids against British towns and cities.

THE NORWEGIAN CAMPAIGN

In contrast to the growing shortage of steel that Britain faced, the War Cabinet concluded in March 1940 that supplies of aluminum were sufficient to meet anticipated needs through the end of 1941, "subject to there

[55] During the Second World War, London consisted of twenty-eight metropolitan boroughs, plus the City of London, that is, the "Square Mile." Draft letter from Leslie Burgin and Walter Elliot to London mayors, Feb. 1940, NA, HLG 51/556. No final copy of this letter exists in this file, but other papers indicate that it was in fact sent.

[56] Nellie Sidney Dark, letter to the editor, *Times* (London), 4 Mar. 1940, 7.

[57] E. Leslie Burgin, "Seventh Monthly Report by the Minister of Supply Covering the Month of February, 1940," 23 Mar. 1940, 10, NA, CAB 68/5/51.

[58] Peter Howlett, *Fighting with Figures: A Statistical Digest of the Second World War* (London: Central Statistical Office, 1995), 6.5.

being no disturbance of existing sources of supply, e.g., Norway."[59] The following month, however, German forces invaded both Norway and Denmark.

This development had enormous implications for Britain, and aluminum was only one of the materials affected. In contrast to the United States and Canada, where paper and cardboard remained plentiful throughout the war, Britain contained few timber resources of its own. British imports of wood pulp from all sources declined by 80 percent after Norway fell, and they remained at that low level for the remainder of the war. Two days after Germany attacked Norway, the Ministry of Supply sent the following message to local authorities: "The War developments of the past few days have thrown into dramatic relief the vital importance of salvaging waste paper, cardboard and rags, to take the place of wood pulp from Scandinavian countries. The saving and collection of these materials for repulping at the mills is now not merely a matter of desirability but one of *national duty*."[60]

Despite this strong rhetoric, the Ministry of Supply did not require local authorities to establish recycling programs.[61] Instead, its initial response was to limit consumption by restricting the amount of paper that publishers of books, magazines, and newspapers could use. It regulated the length of publications, the number of copies produced, and even the amount of space devoted to page margins.[62] Book publishers were allowed to acquire only a third as much paper as they had used in peacetime, and newspaper publishers merely a fifth.[63] In April 1940, the Ministry of Supply announced that a penny daily newspaper could not exceed eight pages in length.[64] Over the following year most newspapers would shrink even further, to just four pages.[65] Shopkeepers had to reduce the amount of paper for wrapping customers' purchases, and

[59] E. Leslie Burgin, "Eighth Monthly Report of the Minister of Supply Covering the Month of March 1940," 20 Apr. 1940, 7, NA, CAB 68/6.

[60] Ministry of Supply, Salvage Circular 17 (11 Apr. 1940), Walsall Local History Centre, 235/1. Emphasis in original.

[61] "Salvage of Waste," *Municipal Review*, May 1940, 108.

[62] Valerie Holman, *Print for Victory: Book Publishing in England, 1939–1945* (London: British Library, 2008), 13–15.

[63] United Kingdom Paper in Wartime, 23 Aug. 1943, NA, BT 28/1085; "The Institute of Public Cleansing: Scottish Centre at Glasgow," *Public Cleansing*, June 1940, 314–15.

[64] Herbert Morrison, "Ninth Monthly Report of the Ministry of Supply Covering the Month of April 1940," 23 May 1940, 9, NA, CAB 68/6.

[65] Draft Note on Imports, Shipping and Supplies Prepared for Mr. Harriman, [Mar. 1941], NA, BT 87/40.

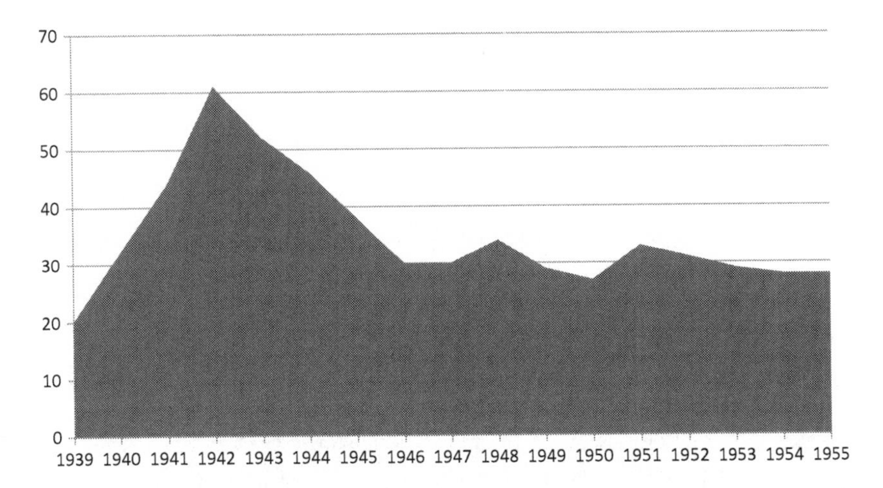

FIGURE 1. Paper recycled as percentage of paper consumed in the United Kingdom (based on weight), 1939–55.
Source: Waste Paper Recovery Association, *Annual Report, 1955* (London: n.p., 1956).

many people began to reuse old envelopes and write on the back of previously used paper. Paper consumption fell sharply, declining from 3.9 million tons in 1939 to 2.5 million tons in 1940. The following year saw a further sharp reduction, and between 1941 and 1945, paper consumption in the United Kingdom averaged just 1.5 million tons a year.[66] Waste paper had provided just 20 percent of the raw material for the manufacture of paper and cardboard at the onset of the war; this figure soon rose to 60 percent.[67] In contrast to this overall decline in paper consumption, the government's use of paper in its own publications increased sharply during wartime, from 40,000 tons in 1939 to 97,328 tons two years later.[68] In addition, the government used vast amounts of paper in the manufacture of munitions.

In peacetime, foreign forests had supplied Britain not only with most of its paper but also with more than 90 percent of the timber it consumed. Wartime restrictions on non-essential construction helped to conserve wood supplies, but large quantities of lumber were needed to build

[66] Waste Paper Recovery Association, *Annual Report, 1955* (London: n.p., 1956).
[67] "Waste Paper in Front Line of Essential Supplies," *WTW*, 14 Sept. 1940, 11.
[68] "Comparative Consumption of Paper and Board in the United Kingdom . . ." [1942], NA, BT 28/1085.

factories, training facilities, and barracks. Another vital use of wood occurred in coal mines, where wooden beams known as pit props were used to prevent shafts from collapsing. In the aftermath of the German occupation of Norway, the *Times* reported that "colliery owners found themselves with all too meagre supplies; yet without pit-props the mines cannot be worked, and without coal industry is paralysed." Britain's landowners had responded admirably to the emergency, the article concluded, through "the sacrifice of woods."[69]

Germany's primary reason for occupying Norway involved neither wood products nor aluminum, but iron. Northern Sweden possesses immense reserves of high-grade iron ore. After it is mined, most of it is shipped by rail to the Norwegian city of Narvik, an ice-free port located about 170 kilometers west of the Swedish city of Kiruna.[70] Herbert Morrison, whom Prime Minister Churchill had just named to lead the Ministry of Supply, observed in May 1940 that the German occupation of Norway had greatly restricted Britain's ability to obtain vital materials: "The loss of Scandinavian supplies is the most important event of the month. Substitutes for these supplies are obtainable, but from more distant sources and often at greater cost. The Swedish ore can be replaced by North African ore, the Swedish steels by United States steels, and the Swedish and Finnish timber by Canadian timber. The Norwegian aluminium is a serious loss and further efforts are being made to increase Canadian supplies. The most bulky supplies that will be lost are the Scandinavian wood pulp and paper. These cannot be entirely replaced even from Canada and the United States and a severe restriction of the use of paper is necessary."[71]

Geoffrey Crowther, the well-informed editor of the *Economist*, offered a more sanguine assessment. The German conquest of Norway, he argued, was essential to Germany, but only an inconvenience to Britain. "Without the Swedish phosphoric ores to mix with her own low-grade ones, Germany cannot produce steel of satisfactory quality. . . . So long as we can import a few million tons of scrap from America, our steel industry can operate, if not at full capacity, at a high proportion of full capacity."[72]

[69] "Woodlands in the War," *Times* (London), 24 Dec. 1940, 5.

[70] Adam Claasen, "Blood and Iron, and der Geist des Atlantiks: Assessing Hitler's Decision to Invade Norway," *Journal of Strategic Studies* 20, no. 3 (Sept. 1997): 71–96; Adam Tooze, *The Wages of Destruction: The Making and Breaking of the Nazi Economy* (New York: Viking, 2006), 380–1.

[71] Herbert Morrison, "Ninth Monthly Report of the Ministry of Supply Covering the Month of April 1940," 23 May 1940, 6, NA, CAB 68/6.

[72] Geoffrey Crowther, *Ways and Means of War* (Oxford: Clarendon Press, 1940), 65–6.

But when the expected scrap shipments from the United States began to dwindle a few months after the German victory in Norway, this optimism turned to anxiety. Even more worrisome was the loss of specialized products from Sweden, one of the few European countries to remain neutral throughout the war. Just a month after Germany invaded Norway, the Air Ministry informed the War Cabinet that the production of aircraft "is being delayed, and men are standing idle, owing to shortage of steel bars, tubes, sheet and stampings. The position has been aggravated by the stoppage of supplies from Sweden, particularly as regards steel tubes and ball bearings."[73]

To obtain vital goods such as these, the Ministry of Supply organized covert missions to sneak them past the German ships that patrolled the Baltic. The first of these was code-named Operation Rubble. It used five Norwegian merchant ships that had been stranded in Gothenburg, Sweden, since the German invasion of Norway. Loaded with 45,000 tons of specialized materials, these ships set sail from Gothenburg in January 1941. After evading several German warships, they arrived safely at Kirkwall in the Orkney Islands in two days.[74] A subsequent mission proved far less successful. Known as Operation Performance, it involved six Norwegian ships carrying "the most valuable cargo which could be obtained in Sweden for the war effort." Although the ships' owners, no doubt under German pressure, tried to prevent them from leaving port, the Swedish Supreme Court denied their request. The ships left Gothenburg on the night of 31 March 1942. The conditions – fog, ice floes, and rough seas – would have been considered dangerous under normal circumstances, but the sailors welcomed them because they would make it easier to evade German naval vessels and planes. Two of the merchant ships soon experienced trouble and returned to port under Swedish escort. Unfortunately for the success of the mission, a full moon shone down on the remaining ships as they left Swedish waters. A short time later, they faced German "destroyers, armed trawlers, a probable submarine and constant air attack." On 1 April the *Storsten* hit a mine and sank fifty kilometers south of Kristiansand, Norway. The *Rigmor* also sank, but the *Lindh* and the *B. P. Newton* made it to Britain. Although only 27 percent of the materials these ships carried reached the United Kingdom, officials in the Ministry of Supply considered the mission a success, for these

[73] "Report for the Month of April (No. 9) by the Air Ministry on Supply and Production," [1940], 1, NA, CAB 68/6.

[74] S. H. Phillips to Andrew Duncan, 30 Jan. 1941, NA, AVIA 11/9.

cargoes were "of extreme value. A large quantity of ball bearings and ball bearing material is available, along with a number of other specialized products which can be regarded as vital necessities."[75]

In the wake of the German advance into Denmark and Norway, many members of the public wrote to the Ministry of Supply to offer suggestions about how various types of salvage could aid the war effort. One correspondent suggested that large quantities of iron could be obtained by melting down garden chains intended to keep people from walking on the grass. "I am quite willing to give mine away," he wrote, "but if it could be made compulsory to part with these chains, all the better."[76] Another idea came from an undertaker, who noted that many tons of brass and iron were "put into the ground daily in Great Britain in the form of handles attached to Coffins." He suggested that if the government outlawed this practice, more metal would be available "to help in this war against tyranny."[77]

Many believed the wartime need for raw materials was so great that the government should require local authorities to establish recycling programs. Yet instead of resorting to compulsion, remarked one critic, the Ministry of Supply had done little besides distributing "a gentle rain of printed matter" to local councils, "urging them to scrutinise and preserve certain of the more obviously desirable portions of their 'refuse,' and emphasising the value and the need of such articles."[78] The thirty-seven-year-old art historian Sir Kenneth Clark, who had become director of the National Gallery seven years earlier, agreed. Speaking at a meeting of the Ministry of Information's policy committee, he argued that if "the campaign for salvage of paper was to be successful, it would be necessary for compulsory powers to be taken to oblige local authorities and others to co-operate."[79]

Harold Judd remained optimistic, at least in public, that a voluntary approach, stimulated by vigorous persuasive messages from the government, would yield high quantities of scrap. Speaking while British forces were trying to oust German troops from Norway, Judd congratulated his

[75] C. R. Wheeler, Report on Operation Performance, 9 Apr. 1942, NA, AVIA 11/23; "MV Storsten," The Clydebuilt Ships Database, The estimate of 27 percent refers to the value, rather than the weight, of these materials. www.clydesite.co.uk/clydebuilt/viewship.asp?id=3956, accessed 29 July 2013.

[76] André Lemaire to Herbert Morrison, received 29 May 1940, NA, POWE 5/56.

[77] Frederick S. Albin to Herbert Morrison, 22 May 1940, NA, POWE 5/56.

[78] C. A. E. Chudleigh, "Apathy and Waste," *Spectator* 164 (23 Feb. 1940): 245.

[79] Ministry of Information, Policy Committee, Minutes, 2 May 1940, NA, INF 1/848.

fellow citizens for having doubled since December the amount of paper they were contributing for salvage. "We want that figure doubled again," he declared, "and with your help it can be done."[80] Chamberlain, already deeply discredited by the failure of his attempts to appease Hitler, lost what little remaining support he had when British forces proved unable to repel the German attack on Norway. On 10 May 1940, Chamberlain handed power to a fellow Conservative, Winston Churchill, who formed a coalition government that would rule Britain until after the defeat of Nazi Germany five years later. Under Churchill's leadership, the government adopted recycling policies far more sweeping than anything previously attempted in Britain. Compulsory salvage soon became a reality for local councils and citizens alike, as did Andrew Bonar Law's dream of tearing up the country's iron railings and melting them into weapons.

[80] "'Waste Less and Salve More': Collecting 130,000 Tons of Paper," *Times* (London), 30 Apr. 1940, 3.

3

Britain's Darkest Hour

> Victory will fall to the side that is best able to husband its resources, moral and physical. . . . Every atom of economical national value is supremely important now, because it might be the one atom that helps sway the scales in our favour.
>
> –*Waste Trade World and the Iron and Steel Scrap Review*, 1940[1]

On 10 May 1940, the same day that Churchill became prime minister, Germany launched a massive attack against the Low Countries and France. Although most people had considered such a move only a matter of time on Hitler's part, its occurrence nonetheless proved deeply unsettling to the British, to say nothing of the Belgians, Dutch, and French. The advance of German forces posed not only a military threat to Britain but also a psychological one. As an official in the Ministry of Information put it on 18 May 1940, "We are in for a period of deep depression and urgent measures should be taken to meet it." He suggested emphasizing two themes: the strength of the British Empire, and a call for sacrifice.[2]

Many people in Britain believed that France's well-fortified Maginot Line would halt the German onslaught and assumed, based on the experience of the First World War, that they were at the start of a protracted struggle to assist the French in holding back German forces. Instead of bogging down as in 1914, this time the German invaders succeeded with terrifying efficiency. Employing new *blitzkrieg* tactics, the Wehrmacht

[1] "'Outshots' . . . by the Rag Man," *Waste Trade World and the Iron and Steel Scrap Review* (hereafter *WTW*), 23 Nov. 1940, 2.

[2] D. Cowan to Lord Davidson, 18 May 1940, NA, INF 1/533.

forced the British army in France to retreat to the coastal city of Dunkirk. When it became clear that they had no prospect of overcoming their encirclement, the British made the painful decision to withdraw from France. Although the "evacuation" of Dunkirk prevented the bulk of the British army from being captured or killed, their rapid flight forced them to abandon most of their vehicles and heavy weapons. This left the United Kingdom with a dire shortage of military equipment at the very moment that Germany seemed poised to invade England.

On 14 June, just five weeks after Hitler launched his attack on France, German troops occupied Paris. By this point Nazi Germany controlled much of the wood pulp, iron ore, bauxite, and other raw materials on which Britain's economy – and its munitions production – normally depended. Although the United Kingdom continued to import food and raw materials from its global empire and from other trading partners beyond Europe, German submarines, warships, and floating mines threatened to cut these vital lifelines at the same time that the Luftwaffe was battling the RAF for control of Britain's skies. As Harold Macmillan observed in July 1940, "We are at the most critical point of total war, at which our country is threatened as never before in history."[3]

Under these trying circumstances, attitudes toward recycling shifted dramatically. No longer simply an admirable goal, salvage became an essential part of the war economy. From 1940 to the early months of 1943, the period when U-boats took their greatest toll on Atlantic shipping, recycling supplied the British war machine with millions of tons of raw materials that were difficult, if not impossible, to obtain from anywhere else. So dire was the situation that the journal *Nature*, which rarely discussed politics or foreign affairs, published the following words shortly after France fell: "Every resource must be used to resist the advance of the aggressor. There is no room for half-measures; the War is the concern of every one of us." To prevail against Nazi Germany, the unnamed author argued, Britons required a new mindset. "Waste household products which in peace-time were rightly consigned to the flames, the rubbish heap, or the attic, become, in these days of war, essential parts of the national economy. Their usefulness must not be judged by ordinary economic standards; the national need is the

[3] Harold Macmillan to Herbert Morrison, 29 July 1940 (copy), Harold Macmillan Papers, Department of Special Collections and Western Manuscripts, Bodleian Libraries, University of Oxford, MS. Macmillan dep. c.267, fol. 210. Extracts from the Harold Macmillan Archive are reproduced with the kind permission of the Trustees of the HM Book Trust.

determining factor."[4] One month later, in a leading article that called on everyone to recycle, the editors of the *Times* made the same point, reminding readers that "the Minister of Supply, who has to provide the Army with tanks and guns, shells and equipment, wants all the country's old metal and tin cans, waste paper and bones. Give him all these things plentifully and he will, though no magician, transform them into armament and explosives. . . . What has been classed as rubbish is no longer rubbish. . . . Waste is a weakening of the national effort and the war on waste is one in which every citizen can engage."[5] The interest of the *Times* and other publications in promoting the recycling of paper was not solely patriotic; it was also necessary to keep them in business.

HERBERT MORRISON AND THE MINISTRY OF SUPPLY

In the shake-up of the cabinet that Churchill instituted after becoming prime minister, he sacked Leslie Burgin as minister of supply and replaced him with the Labour MP Herbert Morrison.[6] Churchill rewarded Harold Macmillan, one of the Conservative MPs who had supported him in the machinations that brought him to Downing Street, with the second-highest position in the Ministry of Supply: parliamentary secretary.[7]

Morrison's appointment as minister of supply found widespread backing, albeit for conflicting reasons. Some lauded Morrison's advocacy of strong government intervention as a refreshing break from the laissez-faire preferences of his predecessor. Speaking while the Battle of Dunkirk was raging, the Welsh Labour MP Jim Griffiths castigated Chamberlain's government for having treated the supply of vital raw materials with complacency. "The country is beginning to realise," he argued, "that we are engaged in a conflict for our lives. . . . What we are fighting is not merely the German Army, Air Force or Navy; we are fighting a German nation which is completely converted into an armed camp. Every bit of German resources and every bit of the resources of every country they conquer is immediately made part of the war machine. . . . We cannot overcome such an enemy by mid-Victorian capitalism in this country.

[4] "Salvage and the Utilization of Waste," *Nature,* 29 June 1940, 988–9.

[5] "No More Waste," *Times* (London), 29 July 1940, 5.

[6] "Service Ministers Changed," *Times* (London), 13 May 1940, 6; David Howell, "Morrison, Herbert Stanley, Baron Morrison of Lambeth (1888–1965)," in *Oxford Dictionary of National Biography* (Oxford: Oxford University Press, 2004).

[7] H. C. G. Matthew, "Macmillan, (Maurice) Harold, First Earl of Stockton (1894–1986)," in *Oxford Dictionary of National Biography* (Oxford: Oxford University Press, 2004).

We can overcome it only if we mobilise the whole resources of this country into one great national effort."[8]

Other Morrison supporters praised his emphasis on conservation as a welcome return to the frugality of the past. Curiously ignoring the austerity associated with the Great Depression, one observer condemned the "extravagance . . . [that] characterised our modern British life. Thrift has been contemptuously cast down from the honoured position it held in Victorian times, when the 'Waste not want not' maxim commanded widespread respect. . . . Pinching, scraping, making do, and extracting every particle of usefulness from every commodity might be all right for poor and backward peoples, but not for well-off Britons. With our standard of life it seemed, misleadingly, that we really could afford to be extravagant." Although this statement sounds like something that a twenty-first century environmentalist might utter, the writer's call for conservation came not from concern about global resource depletion, pollution, or the destruction of ecosystems, but rather from a fear that German submarines might deprive Britain of food and other vital supplies.[9]

The trade publication *Public Cleansing* praised the government's decision to require local authorities to collect salvage, declaring that "the period of free will is at an end. . . . Appeal and education have had a fair trial. But the truth was that local authorities existed who could not or would not look realistically at their responsibilities, and there were householders who wholly failed to realise that we were at war and confronted with short commons, and possibly famine, in many essential materials." Interestingly using the ends to justify the means (something that would happen with increasing frequency as the war progressed), the editorial suggested that perhaps the British people "had a prejudice against compulsion because it was the German method. That need not blind us to its necessity or to its positive qualities. . . . In Germany coercion is employed as a means of terrorism, but in our country [it] is only resorted to for sound public ends."[10]

UP, HOUSEWIVES!

To signal his determination to make domestic salvage a priority. Morrison established an advisory committee on salvage in late May

[8] *Parliamentary Debates*, Commons, 30 May 1940, vol. 361, col. 706.

[9] "A Great Role – Stopping the Leaks in National Waste," *Public Cleansing*, May 1940, 267.

[10] A. L. Thomson, "From 'May' to 'Must' – Compulsion at Last," *Public Cleansing*, Oct. 1940, 62.

1940. Reflecting the widespread assumption that the home was the responsibility of women, he filled this committee exclusively with female members of the House of Commons. To chair it, he turned to Megan Lloyd George, a Liberal MP who represented the Welsh constituency of Anglesey. Her father, David Lloyd George, had served as minister of munitions in 1915 and 1916 before becoming prime minister.[11] The committee's co-chairs came from Britain's other two major political parties. Janet "Jennie" Adamson (Labour) was the MP for Dartford, a town southeast of London, and Irene Ward (Conservative) represented Wallsend in northeastern England. Anticipating a skeptical reaction from its readers, the *Times* insisted that "the appointment of the committee was in no sense a 'stunt'; it was a perfectly serious move," for Morrison believed that these women "would be able to exercise a particularly helpful influence upon housewives, whose cooperation was so essential to success. . . . He had faith in the capacity of the woman M.P.s and believed they would do a real job of work."[12] Three weeks later Morrison announced that upon the recommendation of this committee, he would soon require communities across Britain to collect a wide range of materials for recycling. Justifying this extraordinary step, he declared that "every piece of paper, every old bone, every piece of scrap metal is a potential bullet against Hitler. We should never fling away a bullet. We must never fling away one piece of scrap that can be salvaged."[13]

In reality, preparations to expand rubbish removal to include salvage collection were already taking shape when Morrison appointed the advisory committee. Less than two weeks after Morrison became minister, Harold Macmillan had advised him that it was "very important to act quickly while the country is in the mood to accept almost any sacrifices of vested interests. The exact details . . . can be developed afterwards. Lord Beaverbrook's rapid action in his sphere [aircraft production] has impressed public opinion. We must act equally quickly."[14] In light of the new strategic situation, recycling quickly became a much bigger

[11] "Collection of Salvage: Women M.P.s' Committee," *Times* (London), 31 May 1940, 3; Kenneth O. Morgan, "George, Lady Megan Arfon Lloyd (1902–1966)," in *Oxford Dictionary of National Biography* (Oxford: Oxford University Press, 2004).

[12] "Arms Drive for Victory," *Times* (London), 1 June 1940, 3.

[13] "The Collection of Salvage," *Times* (London), 24 June 1940, 3.

[14] Harold Macmillan to Herbert Morrison, 19 May 1940 (copy), Harold Macmillan Papers, Department of Special Collections and Western Manuscripts, Bodleian Libraries, University of Oxford, c. 273, fol. 267. Emphasis in original. Extracts from the Harold

priority than it had been during the first few months of the war. On 28 May, one day after British soldiers began their retreat from Dunkirk, one of Britain's top salvage officials asserted that it was no longer sufficient to rely on voluntary efforts: "Sound foundations have now been laid on which compulsory schemes could be successfully built up." Using language that emphasized the diminishing distinction between soldiers and civilians in a time of total war, he suggested that "a beginning might be made with urban districts of 50,000 population and over; others could be 'conscripted' as and when practical and necessary." In normal times, he argued, most people would consider such an order an unacceptable intrusion of central government into local affairs, but few would object to it "in the present emergency. In fact, it would have the effect of further emphasising the Government's determination to do everything humanly possible to prosecute the war." Almost as an afterthought, he added, "It would also secure the maximum tonnage of usable waste materials."[15]

On 18 June 1940, Harold Judd met with officials in the Treasury, who sought guarantees that the costs of salvage collection would be borne locally and not passed along to the central government. "If you limit the Order for populations of 20,000 or over," they told him, "the expectation is that the effect of the Order will be to give a profit to most of the Local Authorities concerned."[16] Just a day later, however, Judd informed the Treasury that the Ministry of Supply had decided to expand the order.[17] Morrison soon directed every local authority in England and Wales with at least 10,000 residents, and those with 5,000 in Scotland, to collect salvageable materials from houses, even if they did not pick up residents' rubbish. In a letter to the mayors of every town and city in Britain, he explained, "The national salvage drive is one of the most important phases of the war effort on the home front; we need certain salvaged materials urgently and the need will become greater still as time goes on."[18]

Macmillan Archive are reproduced with the kind permission of the Trustees of the HM Book Trust. Beaverbrook's activities are discussed later in this book.

[15] Unsigned memorandum [probably from J. C. Dawes] to Harold Judd, 28 May 1940 (copy), NA, HLG 51/5.

[16] E. B. B. Speed to Barnes, 18 June 1940, NA, HLG 51/5.

[17] Harold Judd to John Maude, 19 June 1940, NA, HLG 51/5. Sir John Maude (1883–1963) was secretary to the Ministry of Health from 1940 to 1945. See "Sir John Maude," *Times* (London), 7 Feb. 1963, 14.

[18] Herbert Morrison to the mayors of Britain, 19 July 1940, Shakespeare Centre Library and Archive, Stratford-upon-Avon, BRR 55/14/31/6; "Salvage Schemes to Be Compulsory," *Times* (London), 22 July 1940, 2.

These new regulations did not initially apply to Northern Ireland, however. Because of its unusual and contested status, Morrison decided not declined to require local authorities in the province to establish recycling programs when he ordered the rest of the United Kingdom to do so in summer 1940. Six months later, however, the Ministry of Supply announced plans to make salvage compulsory in Northern Ireland.[19] The rest of Ireland, legislatively independent from the United Kingdom but not yet a republic, remained officially neutral throughout the Second World War. Despite this stance, considerable trade occurred between Ireland and Britain during the war, and the British Iron and Steel Corporation purchased significant quantities of iron scrap from firms in Éire (the term the British government then insisted upon in reference to Ireland).[20]

In his efforts to expand salvage collections, Morrison relied on women not only for policy advice and political cover, but also as volunteers who would do much of the labor required to clean, collect, and sort thousands of tons of recyclable materials each week. Although the government's decision to look to women as salvage workers was certainly influenced by economic and workforce constraints, other factors also played an important role. An interesting insight into concerns about public morale can be found in a note that the economist Roy Harrod sent to the physicist and Churchill confidant Frederick Lindemann in March 1940: "While the working classes are thoroughly patriotic about the war, they do not trust the employers or government an inch."[21]

Those in charge of shaping public opinion viewed women as particularly susceptible to anxiety. In June 1940, a civil servant in the Ministry of Information told colleagues that "the large number of women already absorbed in national service are probably strong points of moral [*sic*], but unoccupied women are notoriously weak ones."[22] Another official recommended that "the quickest and easiest way of quieting fears and worries and creating determination and courage is to give the people

[19] "Salvaging Waste Will Be Compulsory in Ulster," *Public Cleansing*, Jan. 1941, 156.

[20] Bryce Evans, *Ireland during the Second World War: Farewell to Plato's Cave* (Manchester: Manchester University Press, 2014), 36–7.

[21] Roy Harrod to Frederick Lindemann, 4 Mar. 1940 (copy), Cherwell Papers, Nuffield College Library, Oxford, F.395/2. Extracts from the Cherwell Papers are reproduced with the kind permission of the Cabinet Office Historical and Records Section. On Lindemann, see Robert Blake, "Lindemann, Frederick Alexander, Viscount Cherwell (1886–1957)," in *Oxford Dictionary of National Biography* (Oxford: Oxford University Press, 2004).

[22] "Report of Planning Committee on a Home Moral [*sic*] Campaign," 21 June 1940, NA, INF 1/849.

something to do – preferably something that will be genuinely useful participation in the war." He went on to suggest a number of specific actions that the public could take, including "various volunteer efforts to collect and put into transportable form metal, old paper, etc."[23]

On 28 July 1940, Morrison delivered a radio broadcast in which he told a nationwide audience that salvage meant "turning our raw materials into war materials." For that to happen, old habits of wastefulness had to be broken. "The fact is, you know, that we in Britain had rather a high old time before the war. Out of the back door of every house in the land, big, medium and small, there used to flow a steady stream of really valuable material mixed up with the rubbish. Nobody thought about it. But now we have to think about it. We have to make the most of what we've got. Every scrap of useful material that we already have in this country must be saved and used again." Addressing an issue that was of grave concern to officials concerned with civilian morale, especially that of women who did not work outside their homes, Morrison said, "I know that many people think that the Government ought to find something active and useful for everyone to do. Some of you feel that your capacity and wish to help are being wasted. Well, here's something that every single one of you can have a hand in, man, woman and child."[24]

Although salvage was a vital national priority, Morrison believed that central government lacked the capacity to collect it. Doing so, he noted, would require "another large army complete with transport and equipment. The people to do it are the local Councils, with the rag-and-bone dealers and various organisations like the Boy Scouts and others helping in their own way." Morrison asserted that most local authorities were actively engaged in salvage efforts, but he insisted that many could do more. To the small number of localities where nothing was being done, he issued a warning: he would not hesitate to issue penalties and prosecute laggard councils if they failed to act. Morrison noted that his talk was the start of a nationwide push to promote the recovery of "every scrap of waste. You will see posters, films and advertisements in the newspapers. You will hear talks on the wireless, and your old pal Syd Walker [an actor famous for his Cockney accent and enormous necktie] will be seeing it

[23] John Rodgers to Lord Davidson, 22 May 1940, NA, INF 1/533. Emphasis in original.

[24] Herbert Morrison, transcript of radio broadcast, 28 July 1940, Trades Union Congress Collection, Modern Records Centre, Warwick University, MSS. 292/557.61/1. Morrison's speech received extensive press coverage. See, for example, "National Salvage Campaign Launched: 'Up, Housewives, and at 'Em,'" *Manchester Guardian*, 29 July 1940, 6.

through in his role of the wandering junk man, which fits in very well." The minister concluded his message by asserting,

> This is a job for us all, but particularly it's a job for the women. In the Salvage Army the whole family is in the ranks, but Mother must lead the way. Here is a chance for you housewives to get right into the fight for freedom and to stay in it until we have wiped these Nazis and every-thing [*sic*] they stand for right off the face of the globe.
>
> I'm going to give you a new slogan for this campaign – my last word to you; "Up, housewives, and at 'em."[25]

In the days that followed, the Ministry of Supply published a series of newspaper advertisements that vividly portrayed the complementarities in wartime of the home front and the battlefield, civilians and warriors, men and women, and civil and military articles. One of these advertisements, addressed "to all careful housewives," contained three separate images. In the first, a soldier with a rifle slung over his shoulder told an elderly woman, "Paper! It means more Ammunition to me." In the second, a sailor told a middle-aged woman, "Metal! It means more Guns to me." The final image depicted an RAF pilot in a flying helmet, who told a young woman, "Bones! They mean more Planes to me." Another advertisement in the series depicted angry housewives literally flinging scrap at the recoiling figures of Adolf Hitler, Joseph Goebbels, and Hermann Göring.[26]

On 1 August 1940, four days after Morrison's radio address, the Women MPs' Salvage Advisory Committee held a press conference – to which only female journalists were invited. According to a press release that the Ministry of Supply issued in connection with this meeting, women could aid the war effort in a vital way by recycling the items that they discarded from their homes. The release attributed to Megan Lloyd George the following words: "Women in ever-increasing numbers are turning out munitions of all kinds. It is now the duty of women to produce the raw material for these munitions." Continuing the focus on women, and emphasizing the benefits to morale that recycling would bring, the statement (which was likely crafted by a government press officer) continued, "They can do their share in winning the war, even

[25] Herbert Morrison, transcript of radio broadcast, 28 July 1940, Trades Union Congress Collection, Modern Records Centre, Warwick University, MSS. 292/557.61/1. Walker died at age fifty-eight after undergoing surgery for appendicitis. See "Mr. Syd Walker," *Times* (London), 15 Jan. 1945, 6.

[26] "Up Housewives and at 'Em!" *Times* (London), 1 Aug. 1940, 2; "Up Housewives and at 'Em!" *Times* (London), 13 Aug. 1940, 7 (quoted above).

FIGURE 2. "Up Housewives and at 'Em!" Four days after Herbert Morrison, the minister of supply, announced the slogan "Up Housewives and at 'Em," his ministry published this advertisement, which implied that ordinary women could transform household items into weapons that would bring Hitler, Göring, and Goebbels to their knees.
Source: Ministry of Supply advertisement, *Times* (London), 1 Aug. 1940, 2.

though they may have no rank other than that of housewife, no uniform but an apron, no munitions factory but the dust-bin. If every housewife does her war job there, as well as it can be done, the effect will reach far beyond the actual salvage rescued from waste. It will turn into tempered steel our national will to victory." The committee had decided that "the only way to get wholehearted co-operation of housewives through the country" would be through face-to-face conversations. To supply the labor necessary to carry out this campaign across Britain, the committee enlisted the help of a large number of women's groups, including Women's Voluntary Services, the National Federation of Women's Institutes, and the women's committees of labor unions and political parties.[27]

Many women responded enthusiastically to the coordinated appeal from Morrison and Lloyd George. As a resident of Walsall, an industrial town in the West Midlands, explained in August 1940, "I've been trying to get my husband to clear out some of this old junk for years, and of course, like a man, he's always said he would do it some time. . . . Now I've got a good reason for doing the job myself and believe me it will be done properly. I'm going to have a regular 'spring clean' and you can tell the collectors that I shall want a cart all to myself."[28] During the drive, municipal workers did not haul away rubbish; their entire labor was devoted to collecting salvage. This created a considerable amount of inconvenience for residents when the duration of the drive grew from just four days to two full weeks. "The result," complained a local newspaper, "was that the whole of the bins in the borough became completely filled. Practically all of them needed two men, instead of one, to carry them to the collecting vehicles, and in addition there were other containers, including baskets and buckets, at many houses." Although this interruption of rubbish collection no doubt caused considerable inconvenience and foul odors, the drive yielded an impressive quantity of materials, including 75 tons of paper and books, and 195 tons of scrap metal in Walsall alone.[29]

[27] Ministry of Supply, Press Office, Notes for a Press Conference of the Women MPs Salvage Committee, 1 Aug. 1940, Trades Union Congress Collection, Modern Records Centre, Warwick University, MSS. 292/557.61/1. For an insightful analysis of government efforts to mobilize women and households, see Geoffrey G. Field, *Blood, Sweat, and Toil: Remaking the British Working Class, 1939–1945* (Oxford: Oxford University Press, 2011), esp. 138–9.

[28] "Housewives' Ready Response," *Walsall Observer*, 17 Aug. 1940.

[29] "Walsall's Salvage Campaign," *Walsall Observer*, 14 Sept. 1940.

During the summer of 1940, the Ministry of Supply enlisted female volunteers to visit every household in Britain in search of salvageable items. Anticipating the large volume of materials that this campaign would bring forth, officials approached the owners of vacant buildings to ask if they would make them available rent-free as collection points. As salvage poured in, these spaces often proved insufficient. In Birmingham, thirty women, including the wife of the mayor, established collection points at their own homes. One of the advantages of these "salvage depots" was that they obviated the possibility, of which many had complained, that dustmen would waste salvage by mixing it with the rubbish they collected.

The industrialist Sir Andrew Duncan, who succeeded Morrison as head of the Ministry of Supply, decided in December 1940 to issue compulsory collection orders for paper and metal to all urban and borough councils in Northern Ireland with 8,000 residents. As was already the case in Great Britain, the Ministry of Supply required these local authorities to submit detailed monthly reports about the types and quantities of materials that they collected for salvage, and it relied on volunteers to go door-to-door to explain the program and deliver leaflets.[30] The volunteer effort in Northern Ireland was led by the Ulster Unionist politician (and member of the province's parliament) Dehra Parker, who served as chairwoman of the Northern Ireland Salvage Committee. To communicate the salvage message to Northern Ireland, the government commissioned newspaper advertisements and radio broadcasts designed to appeal specifically to residents of this province. The campaign also involved Dorothy Waring, a best-selling novelist and Ulster Unionist who had been an enthusiastic proponent of fascism before the war. In January 1941 Waring delivered a broadcast on the BBC Home Service about the importance of recycling.[31]

[30] Ministry of Information, "Salvage Campaign in Northern Ireland," Dec. 1940, NA, AVIA 22/745; Keith Grieves, "Duncan, Sir Andrew Rae (1884–1952)," in *Oxford Dictionary of National Biography* (Oxford: Oxford University Press, 2004).

[31] R. A. Wilford, "Parker, Dame Dehra (1882–1963)," in *Oxford Dictionary of National Biography* (Oxford: Oxford University Press, 2004); W. E. Parsons to B. Crawter, 9 Jan. 1941; H. Edridge, notes, 20 Jan. 1941; B. Crawter to F. M. Adams, 20 Jan. 1941; F. M. Adams to B. Crawter, 14 Jan. 1941, and 21 Jan. 1941; all in NA, AVIA 22/745. On Waring's politics, see Gordon Gillespie, "Waring [*married name* Harnett], Dorothy Grace [*pseud.* D. Gainsborough Waring] (1891–1977)," in *Oxford Dictionary of National Biography* (Oxford: Oxford University Press, 2004).

The calls for women to assist municipalities with salvage proved a welcome one for many local authorities. By the end of 1940 a shortage of labor and vehicles, compounded by the demands of snow removal during the harsh winter of 1940–41, prompted officials to shift the burden of gathering salvage from wage-earning male rubbish collectors (many of whom had left these jobs and gone into the armed forces) to unpaid female volunteers.[32] To promote women's greater involvement in salvage, the Ministry of Supply organized a national conference in the summer of 1941, to which they invited seven hundred women who served as branch secretaries of women's groups across Britain.[33]

In contrast to recycling in the home – which many people considered part of the traditionally female sphere of domestic cooking, cleaning, and caring for children – most viewed salvage in industry, on farms, and in the armed services as a male responsibility. When women expanded their salvage activities into the public sphere, they faced a considerable amount of sexist backlash from men, who often saw them as incompetent busybodies. In an August 1940 letter to *Waste Trade World and the Iron and Steel Scrap Review*, a scrap merchant from the London borough of Fulham informed fellow members of the trade that a woman working for the local council had recently entered his shop.

> Showing me a leaflet, [she] asked if I would co-operate by putting my waste paper, metal, etc., out for the dustman to collect for salvage!
>
> She was lucky I have a sense of humour. I laughed and told her that for years before the last war, my late father, and more recently myself, had been engaged in buying and selling all manner of salvage as it is now called. Very surprised, she assured me she never knew such businesses existed!
>
> It would appear as if these collectors want educating.[34]

SAUCEPANS INTO SPITFIRES

France surrendered to Germany on 22 June 1940. If the German occupation of Norway had made life more difficult for the British, the fall of France threatened Britain's ability to survive as an independent country.

[32] Ministry of Supply, *Memorandum on Salvage and Recovery* (London: HMSO, 1944), 7.

[33] Ministry of Supply, Salvage Department, Area No. 3, "Schemes for Collecting Salvage from Street Depots," n.d. [ca. June 1941], Shakespeare Centre Library and Archive, Stratford-upon-Avon, BRR 55/14/31/4/1; Ministry of Supply, Salvage Department, Area No. 3, Minutes, 25 June 1941, Shakespeare Centre Library and Archive, Stratford-upon-Avon, BRR 55/14/31/4/1; M. I. Allen and J. H. Codling, circular letter, 10 Nov. 1941, Shakespeare Centre Library and Archive, Stratford-upon-Avon, BRR 55/14/31/3/4.

[34] H. Pickford, letter to the editor, *WTW*, 24 Aug. 1940, 7.

FIGURE 3. Aluminum salvage. In response to Lord Beaverbrook's urgent call for aluminum during the 1940 Battle of Britain, millions of people contributed saucepans and other household items for conversion into weapons. One of those who contributed was this Chelsea Pensioner, who visited the WVS depot at Chelsea in July 1940.
Source: Photo by William Vanderson/Hulton Archive. Courtesy of Getty Images (2665308).

In addition to providing Germany with a launching pad for air attacks and a possible naval invasion of Britain, the German occupation of France prevented Britain from obtaining the aluminum it needed to build airplanes. Until May 1940, Britain had obtained 80 percent of its bauxite (the ore from which aluminum is made) from France and French colonies.[35]

[35] "Precautionary Measure," *WTW*, 20 July 1940: 3.

Faced with a shortage of bauxite at a time of rapidly expanding aircraft production, the British government slashed the consumption of aluminum for non-military purposes from 2,500 tons a month when the war started to just 46 tons in August 1940.[36] Aluminum was essential not only in the manufacture of airplanes, but also of explosives and incendiary weapons. Later in the war, strips of aluminum, known by the codename "Window," were dropped from Allied planes in an effort to confuse German radar.[37]

In addition to conserving aluminum, the government embarked on an ambitious recycling program, led by Max Aitken, first Baron Beaverbrook, a Canadian-born newspaper magnate and a Conservative Party heavyweight whom Churchill had appointed to run the newly created Ministry of Aircraft Production. On 10 July 1940, Beaverbrook issued the following call, which he addressed to "the women of Britain":

> Give us your aluminium. We want it, and we want it now. New and old, of every type and description, and all of it.
>
> We will turn your pots and pans into Spitfires and Hurricanes, Blenheims, and Wellingtons. I ask, therefore, that every one who has pots and pans, kettles, vacuum cleaners, hat-pegs, coat-hangers, shoe-trees, bathroom fittings and household ornaments, cigarette boxes, or any other articles made wholly or in part of aluminium should hand them over at once to the local headquarters of the Women's Voluntary Services. . . .
>
> The need is instant. The call is urgent. Our expectations are high.[38]

Both the general and the trade press responded enthusiastically. According to one estimate that appeared a short time later, 5,000 pots and pans contained enough aluminum to build a fighter plane, and five times that number could be turned into a bomber.[39]

Beaverbrook's decision to rely on Women's Voluntary Services as the focal point for the campaign made a great deal of sense. By the summer of 1941, approximately a million women had joined WVS as volunteers, and the organization employed 132 salaried staff members.[40]

[36] Worswick, "British Raw Material Controls," 37, 38.

[37] Ministry of Production, "Report on Non-Military Sector of the British Economy," 5 Oct. 1942, 44, National Archives and Records Administration, College Park, Md. (hereafter NARA), Record Group 169, PI 29 82, Box 695; K. Scott, "Window of Deceit," *Army Quarterly & Defence Journal* 123, no. 1 (Jan. 1993): 39–42.

[38] "Aluminium for Aircraft," *Times* (London), 10 July 1940, 4; D. George Boyce, "Aitken, William Maxwell, first Baron Beaverbrook (1879–1964)," *Oxford Dictionary of National Biography* (Oxford: Oxford University Press, 2004).

[39] "Aircraft from Pots and Pans," *WTW*, 20 July 1940, 2–3.

[40] Memorandum, July 1942, NA, T 162/856.

Not counting the value of the office space, telephones, and stationery that it provided, the British government was spending over £160,000 a year on WVS, £30,000 of which went to town councils for clerical help that they provided to the group's local branches.[41] The same day that Beaverbrook announced the aluminum drive, the head of WVS, Lady Reading, broadcast a nationwide radio appeal in which she called on the public to contribute to the effort. The response was overwhelming, and donations of all sorts began to pile up at WVS offices almost as soon as she finished speaking.[42] By late August, WVS had gathered a thousand tons of aluminum.[43] The superintendent of Buckingham Palace handed over several saucepans and frying pans, and Princess Elizabeth contributed the toy-sized kitchenware from her dollhouse.[44] Responding to posters and newspaper advertisements that encouraged them to imagine household items transformed into weapons, some tried to specify how their donations would be used. "One old lady," noted the WVS *Bulletin*, "parting with her only hot water bottle, was clear that she wanted it made into a Spitfire, not a Hurricane, because after careful study of the papers, she had decided that they were the best planes."[45]

Wartime salvage often squandered the very resources that it claimed to conserve, a point that was not lost on many critics. Commenting on Beaverbrook's call for people to donate aluminum, one observer complained that people were scrapping articles that remained in perfect working condition, including some that were brand new. This was wasteful of money, energy, and labor, and it did nothing to contribute to the war effort.[46] Another overzealous response came from some disabled veterans of the First World War, who attempted to give up their prosthetic limbs in response to Lord Beaverbrook's plea. Officials insisted that such extreme forms of sacrifice benefited no one, but would-be donors often proved difficult to dissuade.

Despite the tremendous response from the public, members of the scrap trade viewed the public campaign as a counterproductive stunt.

41 B. P. Moore to A. F. James, 10 Sept. 1941, NA, T 162/856.

42 *Western Morning News* (Devon), 12 July 1940, 5; "The Appeal for Aluminium," WVS *Bulletin*, Aug. 1940, 1.

43 "Growing Work of W.V.S.," *Times* (London), 19 Feb. 1941, 2; Graves, *Women in Green: The Story of the W.V.S.*, 58–60.

44 "Aluminium from the Royal Family," *Times* (London), 19 July 1940, 7.

45 "The Appeal for Aluminium," WVS *Bulletin*, Aug. 1940, 1.

46 Mervyn O'Gorman, letter to the editor, *Times* (London), 19 July 1940, 5. Emphasis in original.

In the words of one merchant, "There are thousands of tons of scrap aluminium in the country that is [*sic*] not being used, and to publish an appeal for housewives to produce all the aluminium they can is absurd."[47] Beaverbrook responded by saying that the material in the hands of scrap merchants consisted largely of cast aluminum, whereas the aircraft industry needed rolled aluminum. Unfortunately, noted the *Times*, "many of our correspondents had sacrificed domestic belongings made of cast aluminium, since the original appeal invited gifts of any kind of aluminium goods." Large numbers of people were furious at having been misled. "To replace my saucepans with anything else that will cook on a hot plate will cost me three weeks' income," complained one woman.[48]

Responding to his critics, Beaverbrook declared, "We cannot take the pots and pans from the shops. The cost would be too great for the small results we would achieve. No aluminium has been issued for the manufacture of pots and pans since the war started. And so the stocks in the shops must be low. If we were to requisition these stocks, we should set a very big precedent. And we should be acting contrary to the spirit of this appeal to the people to make a voluntary sacrifice for the national cause."[49] These remarks prompted a great deal of criticism. As one person put it, "Lord Beaverbrook's letter is astounding. . . . At a moment of enormous emergency, is he really deterred by the fear of setting precedents?"[50]

KITCHEN WASTE

In terms of weight, the most important material to be salvaged in wartime Britain came from the nation's kitchens. Throughout most of the Second World War, the Ministry of Supply required nearly three hundred communities, home to roughly fifteen million people, to gather food waste for feeding livestock. In addition, more than a hundred other local authorities did so voluntarily.[51] Some cities and towns developed their own promotional materials. In April 1941, for example, Bath issued a leaflet that

47 "Merchants' Stocks of Aluminium," *Times* (London), 12 July 1940, 2.
48 "The Appeal for Aluminium," *Times* (London), 18 July 1940, 2.
49 Lord Beaverbrook, letter to the editor, *Times* (London), 13 July 1940, 5.
50 "The Appeal for Aluminium," *Times* (London), 16 July 1940, 2.
51 G. B. Hutchings, memorandum, 22 Dec. 1943, NA, BT 258/406.

urged residents to save all their kitchen waste so it could be fed to pigs: "Remember, every scrap of foodstuff saved is a blow to Hitler's U-boats which are out to starve us. Here is your chance to beat the enemy in your kitchen. Put your reply to Hitler's threat in the waste food bin." In addition to promoting the collection of kitchen waste, the leaflet called on residents to hand over paper, cardboard, railings, and even old razor blades. Its rhetoric, though intended to inspire, no doubt struck many as trite: "HITLER & MUSSO, those prospective candidates for the World's Scrap Heap, are losing the war. Make no mistake about it. Rub it in by joining the fighting forces with a Spring Offensive in your own home. Store it up for the Hun!"[52]

Kitchen waste provided more than food for pigs and chickens. Animal bones, for instance, had important industrial uses. By the summer of 1940, one South London factory was processing 70 lorries of bones each day. According to a contemporary report, "The bones are sorted, graded, crushed, washed, polished, steamed, milled, and out of them emerge glue, glycerine, fertiliser, and feeding stuffs. . . . Eighty per cent. of the factory's output now goes directly to war purposes."[53]

The collection of waste food posed unique problems. Because of its potential to attract rats and insects, and because it quickly began to smell, food waste – although sometimes collected for salvage by municipal dustmen – was generally deposited in special bins located on street corners, which were to be emptied frequently. Unfortunately, the nation's steel shortage meant that not enough of these covered bins were available. To help address this issue, in 1941 the Ministry of Supply issued a priority allocation of four hundred tons of steel for the construction of collection bins for food scraps.[54]

Like other kitchen scraps, bones quickly became unpleasant during warm weather. To avoid this problem, the government suggested that people dry their used cooking bones by placing them in front of their fireplaces or in their ovens after they finished cooking. "Bones so dried, *not burnt*, will not be damaged for industrial use." A further problem was that even after drying, bones had to be stored separate from other

[52] *City of Bath Salvage Scheme* (Bath: n.p., n.d. [1941]), copy in Shakespeare Centre Library and Archive, Stratford-upon-Avon, BRR 55/14/31/4/1.

[53] "Bones for War Purposes," *WTW*, 17 Aug. 1940, 3.

[54] Ministry of Supply, Salvage Department, "Conference of Honorary District Advisers Held at the Town Hall, Leicester, on Saturday, 20th. September, 1941," Shakespeare Centre Library and Archive, Stratford-upon-Avon, BRR 55/14/31/4/1.

FIGURE 4. Salvaging kitchen waste. This 1942 photograph records one stage of the labor-intensive process required to save, sort, collect, transport, and process food scraps so they could be fed to pigs and other livestock. The salvage of kitchen waste added considerably to wartime food production, but it also helped to spread foot-and-mouth disease from one animal to another. *Source:* Ministry of Information Second World War Official Collection. Courtesy of Imperial War Museum, © IWM (D 7577).

kitchen wastes. Finally, dogs had to be prevented from stealing bones left in bins.[55]

The most serious problem that accompanied the recycling of kitchen scraps was not unpleasant odors but disease. In the midst of the intensive push to persuade people to save their food waste for use as animal feed,

[55] Ministry of Supply, Salvage Circular No. 74 (25 Mar. 1942), Records of the National Federation of Women's Institutes, Special Collections and Archives, London School of Economics, 5FWI/A/3/073. Emphasis in original.

experts traced an epizootic of foot-and-mouth disease to this very source. The outbreak could not have come at a worse time because Britain faced a looming shortage of food for both human and animal consumption. Despite an intensive "ploughing up" campaign that was greatly expanding the amount of land under cultivation, shrinking imports of grain led the Ministry of Agriculture and Fisheries to warn farmers in spring 1941 that they would be able to "obtain only one-sixth of their pre-war requirements" of cereals to feed livestock.[56] Little meat was arriving from abroad; if domestic animals had to be culled because they could not be fed or because they were sick, the nation would face even more serious shortages of calories and protein than it did at present.

Although foot-and-mouth disease does not affect people, it is often fatal to pigs, sheep, and cattle. Moreover, the virus spreads extremely easily, often moving from one farm to another on contaminated vehicles or clothing.[57] In response to the crisis, the Agricultural Research Council convened an emergency conference of experts to discuss steps they might take to prevent the transmission of this and other infections, such as swine fever and trichinosis, "by the distribution of untreated swill collected from household, hotel and army sources." Among those in attendance were three officials from the Salvage Department, including J. C. Dawes.[58]

Meeting with a dozen regional salvage leaders a short time later, Dawes urged them to intensify their efforts to collect kitchen waste. Many local councils had set up collections, he pointed out, but over 600 had not yet done so. At the same time, he warned them, in confidence, of the threat from foot-and-mouth disease, and he emphasized the importance of sterilizing swill.[59] A 1932 regulation stipulated that before it could be fed to animals, swill had to be exposed to boiling temperatures for an hour, the period thought sufficient to destroy infectious organisms. At the time of the disease outbreak, however, only a tenth of the food

[56] Ministry of Supply, Salvage Circular 52 (24 May 1941), East Sussex Record Office, Brighton, C/C/55/132.

[57] Ministry of Agriculture and Fisheries, Foot-and-Mouth (Disinfection of Road Vehicles) Order of 1941 (6071), Shakespeare Centre Library and Archive, Stratford-upon-Avon, BRR 55/14/31/4/1.

[58] Agricultural Research Council, Report of the Conference on Methods of Disinfecting Swill, with Special Reference to Foot-and-Mouth Disease," 8 Apr. 1941, Shakespeare Centre Library and Archive, Stratford-upon-Avon, BRR 55/14/31/4/1.

[59] Ministry of Supply, Salvage Department, District No. 3, Notes of a Meeting in Birmingham, 28 May 1941, Shakespeare Centre Library and Archive, Stratford-upon-Avon, BRR 55/14/31/4/1.

waste collected in Britain was sterilized before being sent to farmers. The Ministry of Supply soon ordered twenty additional concentrator units, but it told local authorities that they would have to pay for them.[60] Not surprisingly, given this half-hearted effort to sterlize swill, the disease continued to spread. After rising sharply during the first part of the war, cases of foot and mouth peaked in 1942 and declined rapidly thereafter. Collections of kitchen waste rose dramatically throughout the episode, growing from 59,000 tons in 1940 to 436,000 tons in 1944.[61] Even at its worst, however, the wartime occurrence of foot-and-mouth disease affected only a fifth as many animals as did the extremely serious outbreak that struck Britain in 2001.[62]

The British government required local authorities to continue collecting food waste until 1953, the same year that it stopped rationing most foods.[63] When the minister of agriculture and fisheries, Sir Thomas Dugdale, announced this policy change, he expressed the hope that many communities would nonetheless continue to recycle food scraps, as doing so would lessen Britain's reliance on imported pork and thus reduce the outflow of foreign currency. In making this plea, Dugdale felt it necessary to reassure the public that the recycling of kitchen waste posed no health risks. Although earlier efforts to sterilize the waste after it reached farms had led to the transmission of disease, he claimed that "no outbreak of animal disease has ever been attributed to treated waste sterilised in a central plant."[64] Decades later, scientists would discover that the use of animal byproducts in agricultural feedstuffs threatens the health not only of livestock, but also of human beings. Heat sterilization cannot prevent the spread of deadly prion diseases such as bovine spongiform encephalitis (BSE), which affected large numbers of British cattle in the 1990s, or Creutzfeldt-Jakob Disease (CJD), a human form of the illness. For this reason, many countries, including the United Kingdom, now prohibit the feeding of animal byproducts to livestock.[65]

[60] Ministry of Supply, Salvage Department, Minutes of a Conference in Leicester, 20 Sept. 1941, Shakespeare Centre Library and Archive, Stratford-upon-Avon, BRR 55/14/31/4/1.

[61] Board of Trade, Salvage Circular 130 (8 Oct. 1947), Ulverston Urban District Council Records, Cumbria Records Office, Barrow, BSUD/U/S/Box 10. Other sources suggest much higher figures. For an estimate of the total wartime salvage of kitchen waste, see Table 5.

[62] The 2001 outbreak reached 2,000 farms and prompted massive culls of potentially infected livestock. See O. O. Peiso et al., "A Review of Exotic Animal Disease in Great Britain and in Scotland Specifically between 1938 and 2007," *PLoS ONE* 6, no. 7 (July 2011): e22066.

[63] H. E. Lambert, memorandum, 12 Dec. 1953, NA, HLG 102/95.

[64] Thomas Dugdale, memorandum, 20 Oct. 1952, NA, HLG 102/95.

[65] Rosalind M. Ridley and Harry F. Baker, *Fatal Protein: The Story of CJD, BSE, and Other Prion Diseases* (New York: Oxford University Press, 1998); Great Britain, *The Introduction of the Ban on Swill Feeding: 1st Report* (London: TSO, 2007).

4

Private Enterprise and the Public Good

> It may be profitable and necessary, or necessary whether it pays or not, to do many things in the interests of national survival and victory which in normal times would not deserve a moment's consideration.
>
> –*Public Cleansing*, 1939[1]

As the war intensified, the government's efforts to maximize the use of recycled materials sparked debates that revealed conflicting visions about the compatibility of patriotism and profits. The war proved quite profitable for medium and large scrap merchants, but many rag-and-bone collectors found it devastating. Rather than enlist their help, as it had with the scrap firms, the government initially urged people to give salvageable items to municipal or charitable groups rather than to rag-and-bone dealers. As a consequence, many scavengers saw their usual sources of scraps disappear. Summoned to court for failure to pay his rent, a rag-and-bone man in Manchester complained in September 1940 that the government's decision to dissuade people from giving materials to scavengers was "a bad job for us chaps. You have not been able to get a living these last eight weeks." He asserted that whereas people formerly allowed "tatters" to have their castoffs, they now preferred to give them to their local council.[2]

[1] "The Dustbin's Contents Again Focuses Interest," *Public Cleansing*, Oct. 1939, 44.

[2] *Manchester Evening Chronicle*, 6 Sept. 1940, quoted in Jack Silverman, "Give the Tatter a Square Deal," *Waste Trade World and the Iron and Steel Scrap Review* (hereafter *WTW*), 21 Sept. 1940, 13.

Wartime price controls also affected the financial viability of for-profit salvage collection. Commenting on this in 1940, the Manchester scrap merchant Jack Silverman wrote that he could understand why the government had established maximum prices in wartime – "to keep prices down . . . and to prevent an undue rise in the cost of living." Silverman argued, however, that the existing system of price controls made little sense. "If, as is probable, the maximum price is in the national interest, I say that it is equally in the national interest to have a minimum price. . . . The people who do the main collecting, the rag-and-bone men . . . are so seriously affected by the low prices that many have been forced to seek other forms of employment. The natural result is that collections suffer. This, in itself, should be a sufficient argument for minimum prices."[3]

Despite using a fictional tatter, played by the actor Syd Walker, to promote recycling, government officials had an ambivalent attitude toward scavengers. Harold Judd, the head of the Salvage Department, considered rag-and-bone men to be inefficient, and his deputy J. C. Dawes shared this view. Writing in May 1940, Dawes asserted that "these small traders should not be allowed to interfere in war time with properly organised municipal salvage schemes. Where practicable these men could be employed to assist the Council's schemes but few would be found suitable." At the same time that he dismissed rag-and-bone men, he expressed a similar disdain for the salvage work of volunteers as "often spasmodic and unreliable."[4]

In his work at the Ministry of Health during the interwar period, Dawes had believed that for the collection of refuse and recyclable materials to be efficient and hygienic, it should be carried out by public agencies rather than private firms. Ten years before the start of the Second World War, he had criticized scavengers who picked through rubbish in search of dirty rags. "If industry needs these materials," he asserted in 1929, "arrangements should be made at the place of disposal for complete sterilisation before sale."[5] In his wartime role in the Salvage Department, Judd sought to prohibit rag-and-bone men from skimming off the most valuable items of salvage and leaving the dross for dustmen to collect at public expense. In a June 1940 memo, he informed local authorities that

[3] Jack Silverman, "Why Not Fix Minimum Prices?" *WTW*, 24 Aug. 1940, 5–6.

[4] Unsigned memorandum [almost certainly from J. C. Dawes] to Harold Judd, 28 May 1940 (copy), NA, HLG 51/5.

[5] J. C. Dawes, *Report of an Investigation into the Public Cleansing Service in the Administrative County of London* (London: HMSO, 1929), 60.

his office planned to issue an order "designed to restrict unnecessary overlapping and competing collections which might interfere with the efficient and economic collection and subsequent use of the materials."[6] Judd's staff duly drafted a regulation that would have made it illegal to "collect or acquire" waste materials from people's houses without authorization from the Ministry of Supply or the local authority.[7] Such regulations, if implemented, would have severely restricted the ability of rag-and-bone men to ply their trade.

Judd's proposal to require individuals to give all of their waste to municipal collectors prompted resistance from inside the government as well as from without. Some argued that such a regulation would create logistical and legal problems in places where local officials had yet to establish efficient recycling programs, and others considered it to be an excessive infringement on private enterprise and individual liberty. Addressing the latter concern, an expert in the Treasury Solicitor's Department thought it unacceptable that an individual "could be prosecuted for burning his waste materials or throwing them into the sea or giving them to his neighbour for his neighbour's bonfire."[8] This would soon happen, however. In July 1942, the North London Police Court fined a man five guineas, plus two guineas in costs, for destroying paper in a fire he had lit.[9]

Over the course of the Second World War, officials asked the British people to recycle a bewildering array of articles, including old batteries, keys, blueprints, and linen. To conserve precious coal supplies as Britain entered the second winter of the war, WVS leaders urged their members to salvage dead tree branches for use as heating fuel.[10] Government officials even asked Britons to give up unwanted gramophone records so they could be used in the manufacture of aircraft components.[11] They also called on people to contribute old pillows and feather beds for an unspecified purpose.[12] Then as now, some items proved easier

[6] Ministry of Supply, Salvage Circular 26 (June 1940), Shakespeare Centre Library and Archive, Stratford-upon-Avon, BRR/55/14/31/6.

[7] Statutory Rules and Orders . . . Salvage of Waste Materials, 3rd Draft, 28 June 1940, NA, HLG 51/5.

[8] Colin H. Pearson to Harold Judd, 3 July 1940, NA, HLG 51/5.

[9] "News in Brief," *Times* (London), 21 July 1942, 2.

[10] "Salvage," WVS *Bulletin*, Nov. 1940, 4.

[11] Ministry of Supply, Salvage Circular 71 (25 Feb. 1942), and 105 (26 Aug. 1943), both in Walsall Local History Centre, 235/3.

[12] Ministry of Supply, Salvage Circular 83 (25 July 1942), Walsall Local History Centre, 235/3.

to collect than to utilize effectively. By the middle of the war, local authorities were collecting 15,000 tons of old batteries each year, but they had difficulty finding firms that could make use of them.[13]

In a 1940 report to Churchill, officials in the Ministry of Supply attempted to put a positive gloss on this problem. "Active steps are being taken," it explained, "with a view to clearing certain temporary accumulations in the markets for low-grade non-ferrous scrap, rags, and bottles."[14] Finding a use for these things was hardly a temporary challenge, for little demand existed for them. One of the reasons that many government officials disliked private scrap collection was that it contradicted their frequent claims that all scrap was equally valuable. Nothing could be further from the truth, however. In the case of scrap iron and steel, experts categorized it into two broad types. Destructor scrap consisted of low-density objects such as wire and empty cans, which were sometimes found in incinerators – that is, destructors – after the combustible materials had burned. Black scrap encompassed heavier items like old tools and equipment.[15] At a meeting of assistant honorary district salvage advisers in October 1941, one of them confessed that he had advised local authorities to "bury the tins and endeavour to arouse salvage enthusiasm again by inaugurating drives for the materials which really mattered, e.g., paper, bones, etc."[16] Behind closed doors, government officials knew that some materials were worth more than others, but they refused to say so publicly. The records of a high-level meeting in November 1941 provide an insight into the government's dilemma. The tin cans and household knickknacks that many people had dutifully saved were "generally of very poor quality and hardly worth the cost of collection, but it was psychologically desirable to clear the dumps when local people were being asked to give up their railings."[17] In response to calls the following year that the collection of tin cans be halted, H. Edridge, the deputy controller

[13] Telegram from Raw Materials Department of the Ministry of Supply to the British Raw Materials Mission in Washington, 10 Nov. 1942, NA, AVIA 22/3088.

[14] Ministry of Supply, Memorandum on the Progress of Salvage Work during 1940, n.d. [Aug. 1940], NA, CAB 120/388 (also in PREM 3/378).

[15] Ministry of Supply, Salvage Circular 1 (Nov. 1939), TNA, HLG 51/556. For a discussion of this circular, see George North to J. C. Carr, 23 May 1940, TNA, HLG 102/92.

[16] Ministry of Supply, Salvage Department, Area No. 3, "Minutes of a Meeting in Birmingham, 1 Oct. 1941," 2, Shakespeare Centre Library and Archive, Stratford-upon-Avon, BRR 55/14/31/4/1.

[17] "Note of Meeting Held at Treasury (Room 42 III) on the 7th November, 1941," NA, T 161/1405.

of salvage, insisted that local officials must continue to gather "metal scrap of all description."[18]

Unless the government found a way to prevent people from taking the most valuable materials from the salvage that local authorities were required to collect, municipal collection of recyclables would become impossible to sustain financially. Salvage officials thus called for tough action against "pirates" who took salvage that owners had set out for collection by municipal employees.[19] In the eyes of many officials, those who sought personal profit from household wastes were thieves because they cheated local authorities of revenue, and saboteurs because they discouraged individuals and officials from doing all they could to recycle. At a meeting of regional salvage administrators in April 1942, one participant asserted that piracy was not limited to individuals. To prove his point, he asserted that the Excelsior Waste Collection and Disposal Company

> deliberately cut across the existing collection schemes of local authorities, and it was generally found that they waited until a local authority had given considerable publicity to their salvage scheme and then they made their house-to-house collections immediately before the advertised date of the local authority collections. Much feeling was being created in the district by the activities of this firm and others similar, and local authorities were very insistent in their demands for the Ministry to take some steps to put an end to them.[20]

In response to such actions, the Ministry of War Transport cut off fuel supplies to the haulage firms that this company had hired to transport its materials.[21]

At stake in such cases was a fundamental ideological disagreement between those who believed that total war required strong central planning and those who believed that private enterprise was inherently more efficient. In July 1940, the journal *Waste Trade World and the Iron and Steel Scrap Review* argued that "at this time of crisis the Trade

[18] Ministry of Supply, Salvage Department, District No. 3, notes of a meeting in Birmingham, 29 Apr. 1942, 4, Shakespeare Centre Library and Archive, Stratford-upon-Avon, BRR 55/14/31/4/1.

[19] G. B. Hutchings, "Report on Salvage," 18 May 1942, 11, NA, AVIA 11/22.

[20] Ministry of Supply, Salvage Department, District No. 3, notes of a meeting in Birmingham, 29 Apr. 1942, 2, Shakespeare Centre Library and Archive, Stratford-upon-Avon, BRR 55/14/31/4/1.

[21] Ministry of Supply, Salvage Department, District No. 3, notes of a meeting in Birmingham, 24 June 1942, 1, Shakespeare Centre Library and Archive, Stratford-upon-Avon, BRR 55/14/31/4/2.

and its products have taken on an essential character second only to that of the actual makers of armaments." The article went on to warn that the public failed to understand the industry or respect its expertise: "The issue is nothing less than whether the Waste Trade is to survive, or, on the pretext of national need, to be superseded by the municipal salvage departments. . . . This threat is a very real one. State Socialism in one form or another holds the peoples of the Continent in its grip, and the thing is too fashionable over here already."[22]

The line between profit-driven rag-and-bone men and public-spirited municipal cleansing staff was in fact often blurred. In July 1940, three dustmen employed by the city of Manchester were convicted of stealing pots and pans that residents had donated for the national salvage drive. The presiding magistrate, who fined each man 40 shillings, admonished them that they "must not assume that certain items of scrap which they collected while working for the corporation were to be regarded by them as a perquisite of their work." Yet this was exactly what many municipal refuse collectors throughout Britain had long done. As a dustman convicted in a separate case put it, the practice of rubbish collectors helping themselves to things that others had discarded had "been going on for generations."[23] Although many local authorities banned municipal employees from profiting from the materials they collected on the basis that it was unprofessional and might promote theft, others paid bonuses to reward the additional effort that salvage required.

In August 1940, Jack Silverman, the director of a large scrap firm in Manchester, argued that despite all the publicity devoted to salvage, "the rag-and-bone man, far from benefiting by this publicity, is actually suffering as a result of it. The general impression created has been that it is a patriotic duty of the housewife to hand over all her rags, bones, paper, and scrap metal to the dustman. In consequence, the rag-and-bone man has become looked upon at best as an interloper, and at worst as a criminal come to pinch the household waste from the Corporations [local government bodies]." Silverman noted that people had long placed "rags, bottles, metals, etc., for the rag-and-bone man, yet it is taken for granted to-day by some authorities that, if any waste materials are left out by a housewife, they are intended for the Corporation dustman." In a thinly veiled reference to Beaverbrook's calls for people to donate aluminum

[22] "Should the Trade Close Down?" *WTW*, 20 July 1940, 2.

[23] "Theft of Salvage: Scrap Not a Perquisite of Dustmen," *Manchester Guardian*, 25 July 1940, 10.

pots and pans, Silverman made an additional point: "It is no cause for boasting that a lot of waste has been collected when we know that much of this waste could have been prevented." Many items that "might be valuable as scrap . . . have a far greater value if they save people from buying new things."[24]

H. Gurney, the director of waste disposal for the London borough of Tottenham, took issue with this view. "One cannot help feeling," he commented,

> that Mr. Silverman's particular object is to cast contempt on the salvage activities of Municipal Corporations, and this he does without being at all consistent or logical. For instance, he tells us that the Councils' efforts are puny and ill-conceived, and in the same breath that their activities have put out of action the professional tatter. . . . As a business man, Mr. Silverman should know that a separate and systematic collection from all dwelling-houses of household saleable waste is not an economical proposition. A local authority, however, is able to do this work more or less efficiently, because it utilises its existing services, i.e., the refuse collection organization, the cost of which is paid for by the ratepayers, and all income resulting from the sale of material so collected is credited to the refuse collection service, and thereby reduces the charge on the rates and so benefits to a small extent every household from which waste is collected.

Gurney went on to state that between thirty and forty men worked for Tottenham to sort and pack reclaimed waste materials. Noting that he paid these men trade union rates, Gurney asserted that their "conditions of employment and remuneration" were at least as good, and probably better, than those of tatters.[25]

Silverman argued that municipal garbage collectors were ill suited to collect recyclable materials, both because of lack of expertise and the absence of a personal incentive to prevent the loss of valuable items. Referring to his opponents' argument that dustmen were already visiting most houses weekly, Silverman responded forcefully. "This may be so, but what was the object of these weekly visits? Was it to reclaim waste materials? Emphatically not! It was – with the exception of a handful of Local Authorities with separation plants – for the exact opposite, to destroy those precious waste materials." Silverman reminded his readers that prior to the outbreak of the war every rubbish van in Manchester had displayed the slogan, "Burn your refuse and reduce your rates."[26]

[24] Jack Silverman, "Give the Tatter a Square Deal," *WTW*, 21 Sept. 1940, 13–14.
[25] H. Gurney, letter to the editor, *WTW*, 12 Oct. 1940, 2.
[26] Jack Silverman, "Co-Operate in the National Interest: Welcome, Mr. Bell!" *WTW*, 30 Nov. 1940, 6–8.

In contrast to its initial disparagement of private scavengers, the Ministry of Supply soon adopted a more welcoming tone. When the Salvage of Waste Materials (No. 1) Order took effect in November 1940, it prohibited unauthorized persons from removing salvageable materials without the owner's consent, but it did not bar householders from selling or giving recyclable materials to scavengers.[27] In an undated pamphlet, most likely published in the second half of 1940, the ministry asserted, "Your regular rag-and-bone man can continue to call for the things he usually takes, if you prefer to sell them to him in this way. They will still help the country." The brochure ended with an exhortation. In language that echoed earlier propaganda messages that admonished people to avoid behavior that the Nazi dictator would welcome, the leaflet demanded, "DON'T SAY 'My little bit won't count.' That's what Hitler would like you to feel. . . . You would not throw away a bullet, a bomber part, a soldier's blanket, a pig's dinner – those are just the things your household waste provides."[28]

In late August 1941, the *Economist* reported that the Ministry of Supply had asked it to publish a long letter extolling the importance and growing success of paper salvage. The editor used the opportunity to criticize the government in two respects, noting that the amount of paper allocated to the magazine was so restrictive that it did not have sufficient space to publish the letter, and complaining that government efforts to promote salvage were conducted in a wasteful manner. To support this charge, the *Economist* asserted that the ministry had sent it "no less than nineteen copies of the letter, each in a separate envelope." One who read this article was Viscount Simon, the lord chancellor, who brought it to the next meeting of the War Cabinet's home policy committee and placed it in the official minutes. A terse accompanying note states that the committee "agreed that the attention of the Ministry of Supply should be drawn to the matter."[29] This incident likely served as the inspiration for a cartoon published that month, which suggested that

[27] The Salvage of Waste Materials (No. 1) Order, 1940 was issued as S.R. & O. 1940 No. 1950. See Great Britain, *Statutory Rules and Orders 1940*, vol. 2 (London: HMSO, 1941), 1451.

[28] Ministry of Supply, *Waste Collection is Now Compulsory* ([London]: Ministry of Supply, [n.d., ca. 1940]), copy in Trades Union Congress Collection, Modern Records Centre, Warwick University, MSS. 292/557.61/1. Emphasis in original.

[29] "Shorter Notes," *Economist*, 30 Aug. 1941, 262, quoted in War Cabinet, Home Policy Committee (Legislative Section), "Minutes of a Meeting of the Committee . . . on Tuesday, 16th September, 1941," 4, NA, CAB 75/10.

FIGURE 5. "Anti-Waste Drive." This satirical critique, which suggested that the government's salvage strategy actually led to waste, appeared in a magazine published by a trade union that represented many of Britain's municipal employees. *Source:* Mortimer cartoon in *Local Government Service,* Sept. 1941, 206. By permission of UNISON, the public service union.

efforts to promote recycling might actually consume more resources than they gathered.[30]

Members of the scrap trade pointed out that rag-and-bone men required no public funds for labor, fuel, or equipment. They further

[30] Mortimer, "Anti-Waste Drive," *Local Government Service*, Sept. 1941, 206.

argued that because scavengers' earnings were entirely dependent on the sale of their finds, they were less likely than municipal employees or volunteers to allow valuable materials to be lost during collection and transport. As one scavenger from Manchester put it, "The 'rag-and-bone men' have collected millions of pounds' worth of waste of all kinds from the housewives of this country much more than the local authorities have ever dreamed of, and have not had to use one ounce of petrol to do it."[31] Another writer, who lived in the adjacent city of Salford, challenged local authorities to disclose not only the revenue that they had generated through the sale of salvaged materials, but also their expenses. According to him, some councils expended three times more money collecting salvage than they received for selling it.[32]

Proponents of municipal control countered that rag-and-bone men were unable and unwilling to carry out a comprehensive recycling campaign, and they warned that private scavengers would simply skim off the most remunerative, easiest to transport items. Allowing them to compete with municipal recycling programs would force the municipal programs to operate at a loss. In response to such arguments, members of the scrap trade maintained that if the government wanted to increase the reclamation of particular materials needed in wartime, it should pay more for them.

At a meeting of regional salvage officials in May 1941, Dawes faced considerable criticism about the way in which the Salvage Department collected and used statistics about salvage at the local level. T. B. Crookes, the chief salvage officer for Scotland, argued that local councils should not inflate their salvage statistics by including materials that were not part of the waste stream. Doing otherwise would foster an inaccurate impression among the public. He noted that "when the figures of various Local Authorities are published in the press they indicate an entirely wrong comparative position as it is obviously unfair to compare a return of domestic salvage only with a return containing several hundred tons of say, tram rails."[33] J. H. Codling, who oversaw salvage in the Midlands, agreed. He pointed out that the decision to report the sale of materials by municipally owned gas, water, electricity, or transport services would

[31] Benny Leach, letter to the editor, *WTW*, 17 Aug. 1940, 8.
[32] Arthur Wright, letter to the editor, *WTW*, 17 Aug. 1940, 8.
[33] Ministry of Supply, Salvage Department, "Conference of Honorary District Advisers Held at the Town Hall, Leicester, on Friday, 16th. May 1941," 1–2, Shakespeare Centre Library and Archive, Stratford-upon-Avon, BRR 55/14/31/4/1.

unfairly disadvantage those communities in which utilities were privately owned. In light of Dawes's insistence that it was essential to keep track of all materials that were salvaged in an area, Codling asked why such tallies excluded the items that rag-and-bone men gathered.

Dawes flatly refused to change the way salvage was calculated and reported, despite the fact that he did count the activities of large paper merchants. It is clear that Dawes disliked the unregulated, difficult-to-monitor activities of scavengers. He welcomed salvage activities that he viewed as complementing the work of local government, but he did not want freelancers to compete with it.[34] For the first two years of the war the Ministry of Supply regularly released data about the various types of salvage that local authorities collected. In the autumn of 1941, however, it suddenly stopped issuing these summaries. The reason for this decision remains a mystery, for the government continued to require local officials to submit detailed reports each month about their salvage work for the duration of the war and well after it ended. It seems probable, however, that this dispute, combined with concerns that much of the low-hanging fruit had already been gathered, played a role in the ministry's decision to withhold monthly salvage statistics from the public.

The ideological disagreements between public and private recycling collection never disappeared entirely, but they assumed less importance as the war persisted.[35] By 1942 Britain faced a decline in the workforce of both scavengers and municipal dustmen. Many had entered military service, and others had been directed by the Ministry of Labour into jobs that it considered more important.[36] As a consequence, in February 1942 the Ministry of Supply announced that every street, office, and factory in Britain would soon have a salvage steward.[37] Although they would be given certificates attesting to their role, salvage stewards would not be issued uniforms or paid for their work.[38]

[34] Ministry of Supply, Salvage Department, "Conference of Honorary District Advisers Held at the Town Hall, Leicester, on Friday, 16th. May 1941," 2, Shakespeare Centre Library and Archive, Stratford-upon-Avon, BRR 55/14/31/4/1.

[35] For a comparative perspective that focuses largely on the economic aspects of waste collection and salvage, see Raymond G. Stokes, Roman Köster, and Stephen Sambrook, *The Business of Waste: Great Britain and Germany, 1945 to the Present* (New York: Cambridge University Press, 2013), esp. chpt. 3.

[36] Hutchings, "Report on Salvage," 12.

[37] Ministry of Supply, Salvage Circular 70 (6 Feb. 1942), Walsall Local History Centre, 235/3; Ministry of Supply, *Memorandum on Salvage and Recovery* (London: HMSO, 1944), 4.

[38] "Salvage Stewards: Organizing Collections in Every Street," *Times* (London), 10 Feb. 1942, 2.

CHILDREN AND SALVAGE

In the autumn of 1939 Women's Voluntary Services began the Cog Scheme, which sought to mobilize an army of children to assist with its recycling work in London. "Each child helping in salvage," the organizers later explained, "is a cog in the great wheel working for victory."[39] Although some people expressed doubts about the program in light of "the natural propensities" of children to litter, the Cog Scheme went ahead.[40] By March 1940, the Ministry of Supply claimed that two million children across Britain were collecting materials for the war effort, mostly through their schools, but also through groups such as the Boy Scouts.[41] To encourage even greater participation, the Board of Education called on every teacher in the United Kingdom to become involved.[42] Officials soon began organizing contests in which schools competed to see who could collect the most waste paper.[43] These contests proved incredibly effective. In an effort to expand their success beyond schools, in 1941 the Ministry of Supply prompted several major newspaper and magazine publishers, as well as the country's leading paper makers, to create a new group, the Waste Paper Recovery Association. Most of the prize money for the contests that followed came from this association, but not everyone supported its work.[44] As one Board of Education official complained in 1942, "We find the element of profit and competition which hangs about the Waste Paper Recovery Association very unattractive, and feel very doubtful whether the schools ought to be encouraged to give time to an activity which results so directly in profit to private persons."[45]

Two years after the Cog Scheme began, WVS expanded it to encompass the entire country. To promote this effort, it issued a colorful brochure, *What a Cog Should Know*. Emblazoned with the group's logo of a cogwheel, the publication included an illustration of a girl pushing a cart full of scrap metal toward a line of tanks. A soldier waved to her from

[39] Women's Voluntary Services, "Salvage," 1 Sept. 1941, Shakespeare Centre Library and Archive, Stratford-upon-Avon, BRR 55/14/31/3/3.
[40] "Salvage of Refuse," *Municipal Review*, Dec. 1939, 394.
[41] Ministry of Supply, Salvage Circular 14 (Mar. 1940), Trades Union Congress Collection, Modern Records Centre, Warwick University, MSS. 292/557.61/1.
[42] E. Leslie Burgin, "Eighth Monthly Report of the Minister of Supply covering the Month of March 1940," 20 Apr. 1940, 9, NA, CAB 68/6.
[43] "Schools Salvage Campaign," *Municipal Review*, June 1940, 118.
[44] Ministry of Supply, *Memorandum on Salvage and Recovery* (London: HMSO, 1944), 2, 4.
[45] J. H. Burrows to H. B. Jenkins, 15 Sept. 1942, NA, ED 10/309.

FIGURE 6. Boy Scouts carry paper to salvage depot. Organized by scouting troops, schools, and Women's Voluntary Services, children throughout the United Kingdom played an important role in the collection of paper, books, metals, and other salvageable items.
Source: Ministry of Information Second World War Press Agency Print Collection. Courtesy of Imperial War Museum, © IWM HU 36212.

the open turret of one tank, and she saluted him in return. The accompanying text explained that in addition to providing the material used to build tanks, scrap metal was needed to make many other weapons, including artillery shells and airplanes. "If we did not save our metal scrap in this country," it noted, "we should have to fetch more metal from abroad, and risk seamen's lives in fetching it." The other side of the document depicted a boy and girl shaking hands with a sailor and an airman, while a soldier watched over the group. "All Cogs," the leaflet explained, "should be the Salvage Officers in their homes, and help their mothers get the paper, rags, metal and bones ready for the dustman before collection day."[46] In addition to such leaflets, the government

[46] Women's Voluntary Services, *What a Cog Should Know* ([London]: Women's Voluntary Services, 1941), copy in Royal Voluntary Service Archive & Heritage Collection, Devizes, WRVS/800/Salvage.

issued nearly 200,000 special badges to children who took part in the Cog program. Many additional children were to have received these badges, but wartime shortages caused a suspension of their manufacture.[47] Another way that the group sought to promote salvage was by distributing "The Cog's Song." Its opening line, "There'll always be a dustbin," borrowed unashamedly from the title of the 1939 hit popularized by Vera Lynn, "There'll Always Be an England."[48]

Building on the extensive work of Women's Voluntary Services with children, the government also used schools to assist in the recycling campaign. In 1942 the Directorate of Salvage and Recovery published *The School Salvage Steward's Guide*, which opened with the assertion, "As a Salvage Steward of your School you have an opportunity of playing an important part in the National War Effort." The document went on to explain, "Your duty as a School Salvage Steward is to multiply the amount of salvage collected in your school, to see to it that the boys or girls in your form, or in your House, know the value of their efforts, as well as the value of waste materials – paper, metal, rubber, rags, bones and kitchen waste. Part of your work will be propaganda. Once you have made the story clear that we must not use our shipping to bring to this country the material we have in our homes, we are a step nearer Victory."[49]

The directorate also issued a set of lecture scripts for schools that were simultaneously jocular and macabre. The first talk, "How Salvage of Waste Paper Helps the Country," explained that it was used to make

> many things that are helping us to win the war. Waste paper is employed in making the round cardboard boxes in which shells are packed and for parts of the insides of mines, bombs and cartridges. . . . In every cartridge there is a bullet at one end and an explosive – generally cordite – at the other, and they are separated by a disc of cardboard called a "wad"; one old envelope will make fifty wads for rifle or machine-gun cartridges. So when a pilot of a Spitfire rattles out thousands of shots from his eight machine guns he uses up quite a lot of old envelopes, doesn't he?

To emphasize the material connection between civilian and military articles, the lecture offered several examples designed to appeal particularly to schoolchildren: three comic books could make cardboard cups for two twenty-five-pound artillery shells, and a breakfast cereal box could make

[47] Charles Graves, *Women in Green: The Story of the W.V.S.* (London: Heinemann, 1948), 204.

[48] Women's Voluntary Services, "The Cog's Song," Royal Voluntary Service Archive & Heritage Collection, Devizes, WRVS/800/Salvage.

[49] Ministry of Supply, *The School Salvage Steward's Guide* ([London]: Ministry of Supply, 1942), copy in NA, ED 10/309.

two practice targets. "Twenty-five sheets of examination paper," it added, "make nine aero-engine dust covers. I expect you think that that is the best use for it!" By saving every scrap of paper for salvage, "you will be doing important work in helping our gallant sailors, soldiers and airmen and everyone who is striving to win the war." The booklet's chapter on non-ferrous metals told schoolchildren, "You may help to make bullets for our commandos yourself if you have any old lead toys or soldiers which you have grown too big for.... If you have a small pair of scales at home you can amuse yourself weighing out your bits of brass clock-wheels and old screws, and see how many cartridge cases you can supply for our soldiers."[50]

Many children were so eager to collect recyclables that they devoted part of their lunch period to the activity.[51] Not all of their efforts were legal. Some children took paper and other materials that had been placed in communal receptacles, while others went so far as to steal from scrap merchants. In one case, an exasperated dealer refused to pay for material collected in a school scrap drive because "most of it had been taken from his own scrap yard."[52] Many officials expressed similar frustrations about children stealing salvage – a phenomenon which the television series *Foyle's War* touched on in one episode.[53]

Irritated by the tendency of the Ministry of Supply to treat schoolchildren as a limitless source of unpaid labor, officials in the Board of Education pointed out that they were already doing a great deal to help. In September 1942, one education official asked for a single "comprehensive list of things which the schools can do during the next few months, rather than a series of separate appeals, each involving more than 1,000 separate sheets of paper, with postage etc. to correspond." A week later, another civil servant in the Board of Education declared that the "sporadic and uncorrelated appeals from the Ministry of Supply for the assistance of school children are getting beyond a joke. Were we to issue any more documents on the salvage of paper they would probably find their way to the waste paper basket and in due course make their own contribution to paper salvage."[54]

[50] Ministry of Supply, *Salvage: Lectures to Schools and Test Papers (With Answers)* ([London]: Ministry of Supply, 1942), 2–3, 7, copy in NA, AVIA 22/3088.

[51] E. O. Leadlay, "Waste Paper Goes to War," *WTW*, 18 July 1942, 22.

[52] Ministry of Supply, Salvage Department, District No. 3, notes of a meeting in Birmingham, 25 Feb. 1942, 4, Shakespeare Centre Library and Archive, Stratford-upon-Avon, BRR 55/14/31/4/2; note by H. B. Jenkins, 23 Apr. 1942, NA, ED 11/269.

[53] Hutchings, "Report on Salvage," 11–12; *Foyle's War*, "War Games," 3 Nov. 2003.

[54] J. H. Burrows to H. B. Jenkins, 15 Sept. 1942, NA, ED 10/309; M. G. H. to H. B. Jenkins, 22 Sept. 1942, NA, ED 10/309.

PROPAGANDA

To encourage people to donate both their property and their labor to the salvage program, the British government used an extensive array of propaganda techniques. Planning for this effort began in earnest in January 1940 when Judd and Dawes met with John Hilton, professor of industrial relations at Cambridge University and director of home publicity at the Ministry of Information, and with R. A. Bevan, one of the top advertising executives in Britain.[55] A series of carefully choreographed public events soon followed. On 28 February the Borough of Tottenham, a leader in municipal salvage, hosted a visit of the minister of supply, officials from the Salvage Department, approximately fifty MPs, and councilors from several London boroughs. The centerpiece of their visit was Tottenham's salvage separation plant. In this state-of-the-art facility, reported the *Times*, "every ounce of refuse collected is turned to profitable use. . . . Rubbish is sorted by men who stand on either side of a long conveyor belt and pick out the trifles allotted to them – one tins, another paper, and so on. Tins and old buckets are pressed into billets for delivery to blast furnaces and conversion into pig-iron by a machine which looks like a fearsome rack which has survived from more barbarous times." Following their visit to the salvage plant, the group enjoyed a meal at Tottenham Town Hall. As a sign of the borough's commitment to adaptive reuse, the organizers printed the luncheon menus on used envelopes.[56] Tottenham gained further attention when Queen Mary and Queen Elizabeth visited its salvage plant on a warm July day in 1940.[57] Judging from the wrinkled nostrils that appear in some of the photographs taken that day, the summer heat made for a fragrant experience.

In addition to high-profile visits, the British government employed a wide variety of other techniques to promote recycling. These efforts included short films such as *Feed the Furnaces*, which asked viewers to donate

> all the old scrap metal articles you can possibly collect. Think of all these cumbersome monstrosities rusting away in the odd corners of your house and garden. That old lampshade that you've been meaning to throw away for the last 10 years. That rusty old lawn-mower that you fall over each time you go in the shed; those

[55] Notes of a conference on salvage publicity, 2 Jan. 1940, NA, HLG 51/556.
[56] "War on Waste," *Times* (London), 29 Feb. 1940, 5.
[57] "10,000 Cheer: The Queen at Tottenham Yesterday," *Tottenham and Edmonton Weekly Herald*, 12 July 1940, 1.

FIGURE 7. Royal visit to municipal salvage works. On 11 July 1940 Queen Elizabeth (later known as the Queen Mother) visited the municipal salvage works in Tottenham to express royal support for recycling. Her mother-in-law, Queen Mary, accompanied her to the works.
Source: Tottenham Public Library and Museum Local Collection. Courtesy of Bruce Castle Museum (Haringey Culture, Libraries & Learning), 933 CLE, Idbcm: 2005. 204.

flat-irons; the old kitchen range; and the children's old toys . . . they're all useful to throw into the war against Hitler. . . . So, whether you are a factory owner, or a farmer, or a householder, bring out your scrap, and feed the furnaces.[58]

One of the most extraordinary aspects of Britain's wartime recycling program was the government's attempt to convince people to look on rubbish as a strategic resource. The Ministry of Information as well as the Ministry of Supply devoted massive resources to promote recycling. By early 1944, 136 people worked in the publicity and campaigns branch of the Ministry of Supply. Half of them were assigned wholly to the Directorate of Salvage and Recovery.[59] Employing the famous cartoonist

[58] Script for *Feed the Furnaces*, n.d., NA, INF 6/439. The Ministry of Information also released a shorter version of this film under the title *Any Old Iron*.

[59] "Public Relations Staffs," *Times* (London), 4 Feb. 1944, 8.

Fougasse (Cyril Kenneth Bird, the art editor for *Punch* from 1937 to 1949) as well as other artists, the ministry disseminated this message in newspapers, magazines, on walls, and in public transport. Despite the lighthearted and often comedic approach of this campaign, its subject was deadly serious.

In contrast to the payment of taxes and the purchase of war bonds – actions that supported the war effort but which had more abstract consequences – officials encouraged people to imagine familiar items of domestic life transformed into bullets and bombs. Herbert Morrison had explored this theme briefly in his July 1940 radio address to housewives, asserting that "old love letters can be turned into cartridge wads, meat bones into explosives, tin cans into tanks, and garden tools into guns."[60] Readers of the *Bulletin*, a magazine that WVS published for its members, later learned that twenty-four old keys could make a hand grenade and that a chop bone yielded "enough glycerine to make cordite for two cartridges."[61] Another publication suggested that seven inches of lead pipe could be used to make a hundred machine-gun bullets, four burnt-out light bulbs would yield enough brass for a rifle cartridge, and that a brass candlestick could be made into four cartridge cases for a 20 mm cannon.[62] Even pianos could be used to make munitions. The Ministry of Supply estimated that the average piano contained 160 pounds of cast iron, 40 pounds of steel, 5.5 pounds of copper, and 3.5 pounds of brass – enough to make seven anti-aircraft shells.[63]

In March 1940, the mayor of Westminster opened an exhibition designed to provide visual examples of the weapons that could be made from waste. The exhibition was housed in Bond Street, one of the most fashionable shopping streets in the West End of London. Speaking at its launch, Harold Judd, the controller of salvage, declared that the nation had reached a turning point in its campaign to put waste to use in the war effort.[64] Perhaps inspired by the municipal example, a few months later the Ministry of Supply opened its own salvage exhibition in an even

60 "Salvage of Waste: Minister's Appeal for Greater Efforts," *Times* (London), 27 July 1940, 2.

61 "Salvage Facts," WVS *Bulletin*, Dec. 1942, 5.

62 "Public Salvage a Vital Necessity,"*Public Cleansing and Salvage*, Sept. 1943, 54. Formerly called *Public Cleansing*, the journal added salvage to its title in July 1942.

63 James Forbes, ed., *The Municipal Year Book and Encyclopædia of Local Government Administration, 1944* (London: Municipal Journal, 1944), 280.

64 "Campaign against Waste," *Times* (London), 6 Mar. 1940, 5.

more prominent location: Charing Cross Underground Station. "The connexion between old iron and guns," commented the *Times* in a glowing review of the Charing Cross exhibition, "is clearly brought out, and a well chosen series of photographs traces the progress of the raw material from the scrap heap to the railway, on through the works, to emerge as ingots, and finally to the aeroplane, A.R.P. [air raid precautions, i.e., civil defense] trailer, tanks, and ships." Speaking at the opening of the exhibition, Morrison told those present that "the war material that had been won from the scrap and waste material of the homes, the fields, and the kitchens of Britain was helping to bring down Dorniers and Messerschmitts at that moment."[65] The exhibition included a film entitled *From Bedsteads to Battleships*.[66]

At the same time that government officials were working overtime to emphasize the remarkable range of armaments could be made from civilian items, the *Times* published an article that contained a potentially alarming heading: "Old Masters for New Bombers." What followed was not some Swiftian satire about a new source of camouflage pigments and canvas for aircraft manufacture, but an idea for how Britain might find the funds to continue its fight against Germany now that most of the democracies in Europe had been conquered. The article discussed the possibility that some of Britain's most valuable works of art could generate funds for the nation's war effort. "Why not set up an exhibition in America and charge $1 admission? Millions of dollars could be earned to pay for food and munitions."[67] In addition to implying that America's monetary wealth was inversely proportional to its cultural riches, the article reflected the growing sense of anxiety that Britain's reserves of foreign currency would soon be exhausted.

In 1942 the government created yet another exhibition intended to encourage renewed efforts to salvage metal. The title and mascot of the exhibit was "Private Scrap," a name that alluded to both the military purpose of salvage collection and its household origins. According to the *Times*, the exhibition included "a trench mortar, a machine-gun, an automatic rifle, shell cases and bullets, made from salved tin cans, waste paper, bones, and rags." Drawing on earlier propaganda in which British women hurled saucepans at Nazi leaders, Private Scrap invited

65 "'Scrap for Victory': Where the Old Iron Goes," *Times* (London), 16 Aug. 1940, 2.
66 "London Scrap Exhibition," *WTW*, 17 Aug. 1940, 13.
67 "Prevention of Waste: Some Suggestions," *Times* (London), 16 Aug. 1940, 2.

FIGURE 8. "Put Out More Salvage for More Munitions." Books and other sources of old paper played an essential role not only in the manufacture of paper and cardboard but also in the production of many weapons.
Source: Ministry of Supply advertisement, published in the *Times*, 2 July 1943, 3.

"housewives to 'send a Valentine to Hitler.' But the warrior was not getting his seasons mixed: the Valentine he had in mind was the useful tank of that name."[68] As with several other wartime exhibitions, this one

[68] "'Private Scrap' in London: Host at a Salvage Exhibition," *Times* (London), 18 Dec. 1942, 6.

FIGURE 9. Salvage exhibition in Westminster. At the opening of a salvage exhibition in Bond Street on 5 March 1940, the mayor of Westminster, Major Richard Rigg, posed with an anti-aircraft shell while a municipal dustman and salvage collector named Cockerton held an old saucepan that could be transformed into such a weapon. Standing beside them was Harold Judd, who led the Salvage Department in the Ministry of Supply.
Courtesy of Westminster City Archives, WccAcc 152/5/10/8a.

traveled around the country. In February 1943, it opened in Jenner's Department Store in Edinburgh.[69]

Many Britons quickly grew weary of this constant barrage of messages exhorting them to recycle. Their frustration intensified when they saw a contradiction between the official line and what was actually happening in their communities. One such critic was Edgar H. Summers, the owner of a chain of cinemas in the West Midlands. When the town clerk in Walsall asked him to show a short film on paper recycling called *It All Depends on You*, he issued this tart reply:

[69] "Edinburgh's Book Salvage Drive Begins," *Public Cleansing and Salvage*, Feb. 1943, 216.

We are quite willing to cooperate in any matter such as this, and although I am of the opinion we have already shewn [*sic*] this film previously, as we show a propaganda film every week for the Ministry of Information, we will shew it again . . . but I would ask you when this particular film is shewn, to get your committee responsible for the salvage collection, to give instructions that the salvage is collected in accordance with the manner stated by the film, as after we shewed the last film dealing with this matter, and the people were asked to sort the different kinds of salvage, after watching your collectors throw it all together into the dust cart, they themselves did not trouble any further with salvage.[70]

This letter was hardly an anomaly. During the first year in which municipalities were required to collect recyclables, Dawes reported that his department had received "hundreds of well-grounded complaints from people who really want to help, and willingly go to the trouble of making materials available, only to see them pitched into the refuse collection vehicle by the collector."[71] In striking contrast to most published reports, which dutifully and unquestioningly followed the government's instructions regarding salvage, *Public Cleansing* magazine published this sharply critical rejoinder in April 1941 from D. J. W. Robertson:

The Government through its Ministry have now been crying salvage for over a year, and separation by the housewife is stressed as an essential part of the scheme. While none of us take the utterances of government departments more seriously than we can help, I feel rather surprised that Cleansing Superintendents all over the country should have swallowed this pre-separation business so whole-heartedly. Is it really so essential, or am I blind, stupid, or merely not salvage conscious?

For goodness sake let us face plain facts! Before the war, the dustman called at each house and fetched out the bin – generally to the front. With the same time at his disposal, we now expect him to bring the bin, a bundle of paper, some rags, the week's bones, the household scraps for pig feeding, perhaps some scrap metal, a quantity of broken glass, and he is sure to be asked by the housewife to take the tin cans separately, and the empty bottles and jars. How many journeys? Just count them up! . . .

If every one separated all their salvage, the dustmen would need to make on an average three journeys to each house. How is this to be done when we can't even get the men for an adequate service of one journey per house. [*sic*] There are districts in the country working with less than half their pre-war strength, and most departments are considerably reduced in numbers. In these districts, struggling against such difficulties, all this talk of pre-separation is definitely harmful.

[70] Edgar H. Summers to town clerk, 7 Mar. 1941, Walsall Local History Centre, 235/2.

[71] J. C. Dawes, "Notes on the Work of the Ministry of Supply Salvage Department and Municipal Salvage," *Public Cleansing*, Sept. 1941, 12.

The ratepayer sees the dustman toss his cans, bottles and bones into the bin and thence to the collecting vehicle, and he either wastes the department's time with indignant complaints, or he saves no more and burns his paper, rags and bones.

Robertson facetiously suggested that the government might soon issue "Salvage Circular XYZ, making compulsory the . . . salvage [of] string, broken crockery and beer bottle corks in districts of over 2000 population."[72] This prediction, intended as hyperbole, came partly true less than a year later when the Ministry of Supply issued the Salvage of Waste Materials (No. 3) Order, which made it a criminal offense to destroy rags, rope, or string. Violators could be fined £500 and sentenced to as much as two years in prison.[73]

[72] D. J. W. Robertson, "Is Pre-Separation of Salvage Really Practicable?" *Public Cleansing*, Apr. 1941, 244.

[73] The Salvage of Waste Materials (No. 3) Order, which took effect 20 July 1942, was issued as S.R. & O. 1942 No. 1360. See Great Britain, *Statutory Rules and Orders 1942*, vol. 2 (London: HMSO, 1943), 1992; "Save Rags, Rope and String: Fines for Throwing Them Away," *Times* (London), 17 July 1942, 2.

PART II

ALLIANCES

5

Lend-Lease

> No belligerent can survive long unless it possesses heavy industries capable of turning out considerable quantities of iron and steel. . . . But factories and plants are of little value without a continuous and sufficient supply of raw materials – the sinews of war. There are few raw materials which are not vital in wartime.
>
> – John C. deWilde et al., 1939[1]

U.S. officials believed that they could learn a great deal from how the British transformed their economy for war, and efforts to salvage raw materials occupied a central place in these investigations. In the summer of 1940, President Roosevelt asked William J. Donovan to travel to London to ascertain Britain's chances of resisting Nazi Germany.[2] Donovan, whose judgment FDR valued highly, would later lead the Office of Strategic Services (OSS), the predecessor to the Central Intelligence Agency (CIA). British officials, desperate for American aid, handed Donovan a top-secret summary of their nation's munitions production. Significantly, the British report included a discussion of recycling. A copy of the document soon reached Edward R. Stettinius Jr., the former head of U.S. Steel whom Roosevelt had recently asked to help organize U.S. rearmament efforts.[3]

[1] John C. deWilde, James Frederick Green, and Howard J. Trueblood, "Europe's Economic War Potential," *Foreign Policy Reports* 15 (15 Oct. 1939): 178–92, quotation from 178.

[2] On Donovan's mission to Britain, see F. H. Hinsley et al., *British Intelligence in the Second World War: Its Influence on Strategy and Operations* (New York: Cambridge University Press, 1979), I:312.

[3] Arthur Greenwood, "Survey of Britain's War-Time Economic Organisation: Note by the Minister without Portfolio," 29 Aug. 1940, Stettinius Papers, Box 91.

Stettinius asked William Yandell Elliott, a former Rhodes Scholar who chaired the Department of Government at Harvard, to study the feasibility of doing something similar in the United States. After examining the issue, however, Elliott recommended that no steps be taken to enlist the American public in recycling efforts. Doing so would only alarm the populace and "fritter away real patriotic enthusiasm in useless or unrewarding activities." Instead of collecting materials for recycling, he argued, the United States should seek substitute materials and practice conservation.[4] Despite Elliott's skepticism about recycling, Stettinius thought that the idea held promise; in late August he informed the president that he was exploring ways to enlist state and local officials' assistance in collecting scrap rubber and paper throughout the United States.[5]

Even though the United States was not yet at war, Roosevelt took steps to prepare the country for that eventuality. In the summer of 1940, with rearmament expanding rapidly, the American government signalled that it might soon restrict exports of scrap iron and steel. This news provoked fears in London that British mills would be forced to reduce their output of steel, with dire consequences for the nation's war effort. In light of these concerns, the British government sought to exploit hitherto untapped sources of raw materials within the United Kingdom and throughout the empire. Shipments of iron and steel scrap from India to Britain rose enormously, growing in value from just £56,000 in the 1939–40 fiscal year to an annual rate of £3,000,000 during the summer of 1940.[6]

That September the British consul general in New York informed London that the "scrap position here has become extremely difficult." The Roosevelt administration was considering a total embargo on the export of iron and steel scrap, and even if this could be averted, a massive reduction was likely.[7] In the eyes of a well-placed British official in Washington, the reason for the American decision was primarily economic, as a ban on exports would counter the rising price of steel in the United States, which had increased by 50 percent in the year since Germany had invaded Poland. He added, however, that the State Department likely hoped that the export embargo would serve

[4] The quoted words are from a summary of Elliott's memorandum. See Hayden Raynor to Edward R. Stettinius Jr., 8 Aug. 1940, Stettinius Papers, Box 87.

[5] Edward R. Stettinius Jr. to FDR, 29 Aug. 1940, Stettinius Papers, Box 85.

[6] "Exports of Scrap-Iron from India Have Risen," *Times* (London), 2 Oct. 1940, 8.

[7] I. F. L. Elliot to A. K. McCosh, 17 Sept. 1940, NA, POWE 5/58.

"as a further measure of 'warning' against Japan."[8] In late September the Roosevelt administration announced a ban on exports of scrap to Japan effective 16 October.[9]

In November 1940, only weeks after Congress enacted the first peacetime draft in U.S. history, the American people elected Roosevelt to an unprecedented third term as president. With the election over, he began an effort to persuade his skeptical fellow citizens that the United States should play a more direct role in the European conflict. In his famous "garden hose" speech of 17 December, Roosevelt asserted that furnishing aid to Britain was analogous to one neighbor helping another to put out a house fire.[10]

A short time later, he dispatched his close friend and adviser Harry Hopkins on a mission to Britain. On the morning of 6 January 1941, Hopkins boarded a Pan American Clipper in New York and began a perilous and circuitous journey to England, which he reached three days later.[11] The day after he arrived in London, Hopkins met with Churchill at No. 10 Downing Street. In a handwritten letter, Hopkins informed Roosevelt that the prime minister's official residence was "a bit down at the heels" as a result of the ongoing air raids. "Most of the windows are out – workmen over the place repairing the damage. Churchill told me it wouldn't stand a healthy bomb." The prime minister, he told Roosevelt, "several times assured me that he would make every detail of information and opinion available to me and hoped that I would not leave England until I was fully satisfied of the exact state of England's need and the urgent necessity of the exact material assistance Britain requires to win the war."[12] Despite each side's show of transparency, neither was as forthcoming as their alliance might suggest. As a case in point, when Hopkins attended a meeting of the War Cabinet during his visit, they

[8] A. D. M. to H. N. Sporborg, 14 Sept. 1940, NA, CAB 115/123.

[9] "US Embargoes Scrap Iron, Hitting Japan," *New York Times*, 27 Sept. 1940, 1.

[10] Frank L. Kluckhohn, "Aid Plan Outlined," *New York Times*, 18 Dec. 1940, 1.

[11] "Hopkins on His Way to London by Air," *New York Times*, 7 Jan. 1941, 7. On the fascinating life of Harold (Harry) Lloyd Hopkins (1890–1946), see Robert E. Sherwood, *Roosevelt and Hopkins: An Intimate History* (New York: Harper & Brothers, 1948); and David L. Roll, *The Hopkins Touch: Harry Hopkins and the Forging of the Alliance to Defeat Hitler* (New York: Oxford University Press, 2013).

[12] Harry Hopkins, memorandum, 10 Jan. 1941, Harry L. Hopkins Papers, Georgetown University Library Special Collections Research Center, Washington, D.C., Part III, Box 6, Folder 30. Extracts from the Harry L. Hopkins Papers are reproduced with the kind permission of Professor June Hopkins.

saved a crucial item of business – the relationship between the United States and Japan – until he had departed.[13]

Among the many remarkable people Hopkins met in London, none impressed him more than King George VI and his queen consort, Elizabeth (later known as Queen Elizabeth the Queen Mother). In a letter to Roosevelt in which he described his impressions of this pair, Hopkins declared, "If ever two people realized that Britain is fighting for its life it is these two. They realize fully that this conflict is different from the other conflicts in Britain's history and that if Hitler wins they and the British people will be enslaved for years to come." On 30 January, while Hopkins was dining with the royal couple at Buckingham Palace, air raid sirens sounded outside. The king assured Hopkins that this was a regular occurrence and continued eating, so Hopkins followed suit. A short time later, however, a bell began ringing within the palace, and the king announced, "That means we have got to go to the air raid shelter." Hopkins followed the king and queen to the basement, where he saw the emergency quarters where they slept during bombing raids on London. There they resumed their conversation. "When I emphasized the President's great determination to defeat Hitler, his deep conviction that Britain and America had a mutuality of interest in this respect, and that they could depend upon aid from America, they were both very deeply moved."[14] After five weeks in Britain, Hopkins returned to Washington, where Roosevelt was hard at work lobbying Congress to approve a controversial and mammoth foreign-aid proposal: the provision of billions of dollars in Lend-Lease assistance to the United Kingdom.[15]

HARRIMAN

One of the most vocal supporters of the Lend-Lease bill was Harry Hopkins's friend, the investment banker and railroad tycoon W. Averell Harriman. In a short but powerful speech to his fellow alumni at the Yale Club in New York City on 4 February 1941,

[13] David L. Roll, *The Hopkins Touch: Harry Hopkins and the Forging of the Alliance to Defeat Hitler* (New York: Oxford University Press, 2013), 117.

[14] Harry Hopkins, Diary, 30 Jan. 1941, Harry L. Hopkins Papers, Georgetown University Library Special Collections Research Center, Washington, D.C., Part III, Box 6, Folder 30. George VI reigned from 1936 until his death in 1952; Queen Elizabeth the Queen Mother lived another half century and died in 2002 at the age of 102.

[15] Sherwood, *Roosevelt and Hopkins*, 278; Warren F. Kimball, *The Most Unsordid Act: Lend-Lease, 1939–1941* (Baltimore: Johns Hopkins University Press, 1969).

Harriman vigorously endorsed Roosevelt's efforts to persuade the American people to do more to stop Nazi Germany. "Are we willing to face a world dominated by Hitler? If not, we still have time to aid Britain. If we are to aid Britain, let's be practical and grant the President power to make our aid effective. The most fatal error would be half hearted and insufficient help."[16] Four days later the Harvard professor and Roosevelt adviser William Yandell Elliott sent the text of Harriman's speech to the president's administrative assistant, James Rowe. Elliott praised the address, noting that his friend "Bill Batt, Jr. told me the other day that Averill [*sic*] Harriman had made the best speech that he had heard over WOR [radio station in New York City] the other night. . . . It was particularly satisfying to have a business man of Averill's standing come out flat-footed as he did for the Presidential powers in the Lend-Lease Bill, and I am only sorry that the speech didn't get more public attention."[17]

The same day that Harriman spoke at the Yale Club, the Foreign Office cabled Lord Halifax, the British ambassador in Washington, with the question, "Do you consider Mr. Harriman's sympathies to be pro-German? Importance lies in rumours of his appointment as Minister at United States Embassy here." A handwritten note accompanying the outgoing telegram observed, "I imagine we can't do anything to stop the Pres't appointing him here but we might possibly ask Washington['s] comments."[18] These concerns may have stemmed from the fact that several German companies did extensive business with Brown Brothers and Harriman and Co., the bank that Harriman controlled. Such fears proved unwarranted; in fact, some U.S. officials soon came to worry that Harriman was so sympathetic to the British position that he was neglecting America's interests.

[16] "Address by W. A. Harriman on the Occasion of the Annual Dinner of the Yale Alumni University Fund Association, Yale Club, New York, Tuesday Evening, February 4, 1941," Harriman Papers, Box 158, Folder 4; Rudy Abramson, *Spanning the Century: The Life of W. Averell Harriman, 1891–1986* (New York: W. Morrow, 1992).

[17] William Yandell Elliott to James H. Rowe, 8 Feb. 1941, James H. Rowe, Jr. Papers, Franklin D. Roosevelt Library, Hyde Park, N.Y. (hereafter FDR Library), W. Y. Elliott folder. Many people mistakenly initially assumed that Harriman's name was spelled Averill; some of the letters that the British sent him contain an "e" typed over the "i" in an attempt to correct the error.

[18] Foreign Office telegram to Washington, No. 669, 4 Feb. 1941, NA, FO 954/29A; D. J. Dutton, "Wood, Edward Frederick Lindley, First Earl of Halifax (1881–1959)," in *Oxford Dictionary of National Biography* (Oxford: Oxford University Press, 2004).

On 18 February 1941 FDR held a news conference at which he named Harriman his "defense expediter" to Britain.[19] Before sending Harriman to London, FDR gave him a letter of appointment that described him as "my Special Representative, with the rank of Minister, in regard to all matters relating to the facilitation of material aid to the British Empire."[20] From an office attached to the U.S. embassy in Grosvenor Square, Harriman would coordinate the distribution of U.S. assistance. He was to handle all the economic aspects of the "special relationship," and Ambassador John Winant was to manage the politics. Inevitably, of course, there was no way to disentangle them.

On the eve of his departure, Harriman met with Sir Hugh Dowding, who had led the RAF Fighter Command to victory the previous year during the Battle of Britain. Dowding gave Harriman a glowing letter of introduction to carry with him to London. In it, the air chief marshal wrote, "I cannot refrain from letting you know how highly I esteem him. . . . If I call him 'The Man with the Ten Talents,' you will understand what I mean."[21] The next day, British officials stationed in Washington hosted a dinner in Harriman's honor at the Carlton Hotel, two blocks from the White House. Among those present were Arthur Purvis, head of the British Supply Council in North America; the French diplomat and later architect of European integration, Jean Monnet; and the Australian industrialist Sir Clive Baillieu, director of the British Purchasing Commission. The day after the dinner Baillieu wrote to Harriman to tell him how much he had enjoyed meeting him. "My best wishes for every success in your task. I can imagine few which will be more interesting and absorbing than the one entrusted to your care, and England may count herself fortunate that one knowing her ways so well is representing your great country at this critical hour."[22]

[19] "Excerpts from the President's Press Conference of February 18, 1941," Harriman Papers, Box 158, Folder 4.

[20] FDR to Averell Harriman, [6 Mar. 1941], Harriman Papers, Box 158, Folder 5. Harriman's files include a note dated 11 Mar. 1941 indicating that although Roosevelt signed and handed the appointment letter to Harriman on 6 Mar. 1941, the original was undated. The files include another copy of this letter that includes this date, which was likely added by Harriman's staff. See also Sherwood, *Roosevelt and Hopkins*, 269.

[21] H. C. T. Dowding, letter of introduction for Averell Harriman, 26 Feb. 1941, Harriman Papers, Box 158, Folder 4.

[22] Clive Baillieu to Averell Harriman, 28 Feb. 1941, Harriman Papers, Box 158, Folder 4; Simon J. Potter, "Baillieu, Clive Latham, first Baron Baillieu (1889–1967)," in *Oxford Dictionary of National Biography* (Oxford: Oxford University Press, 2004).

U.S. economic assistance came with many conditions. The Roosevelt administration, sensitive to public opinion, required the British to demonstrate that they were using money and materials as efficiently as possible. As part of this project, U.S. officials repeatedly pressured their British counterparts to intensify their salvage activities. In a move no doubt intended to win points with Washington, just as Lend-Lease was clearing its final legislative hurdles the Ministry of Supply announced a major expansion in the number of municipalities required to operate recycling programs. Henceforth all towns in England and Wales that had at least 5,000 residents, and all rural areas of 10,000, would have to collect salvageable materials from residents.[23]

Confident that Congress would approve his Lend-Lease bill, FDR sent Harriman to London on 10 March 1941, one day before it cleared the House of Representatives and Roosevelt signed it into law.[24] Over the next four and a half years, the United States provided approximately $50,000,000,000 in food, raw materials, and weapons to Allied countries. More than half of this total went to Britain.[25]

In a revealing memo that he dictated on his way to Britain, Harriman created an aide-mémoire of some of the conversations he had held before his departure. "Hopkins," he noted,

> does not realize that the reason why his fourteen points request [for aid to Britain] is not being accepted by those in charge in the army and navy is that he has given them no proof that it is better for the joint effort to have the material in the hands of the British than in our own hands. Our people know the use to which they propose to put the material and have no knowledge of what the British will do. Such remarks are made as "We can't take seriously requests that come late in the evening over a bottle of port" which, without mentioning names, obviously refers to evening conversations between Hopkins and Churchill.[26]

As he made his way across the Atlantic, Harriman carried with him a memorandum from Professor Elliott. In a reversal of his earlier arguments

[23] "More Salvage Essential: Paper, Metal, Rags and Bones," *Times* (London), 4 Mar. 1941, 2.

[24] "Harriman Leaves as Special Envoy," *New York Times*, 11 Mar. 1941, 12; "President Signs, Starts War Aid," *New York Times*, 12 Mar. 1941, 1.

[25] In total, the United States provided about $30,000,000,000 in Lend-Lease aid to Britain, some of which Britain repaid in the form of goods and services that it provided during the war to the United States. Britain paid off the remainder, which was largely used to finance immediate postwar construction, in 2006. See Charles P. Kindleberger, *A Financial History of Western Europe*, 2d ed. (New York: Oxford University Press, 1993), 413–16; and Finlo Rohrer, "What's a Little Debt between Friends?" *BBC News Magazine*, 10 May 2006, http://news.bbc.co.uk/2/hi/uk_news/magazine/4757181.stm, accessed 29 Oct. 2014.

[26] Averell Harriman, memorandum, 11 Mar. 1941, Harriman Papers, Box 158, Folder 5.

that American officials should not promote recycling among the public, he told Harriman that the United States had much to learn from British "methods of collection of scrap." Elliott also urged Harriman to take a tough line with British officials in pressing U.S. interests: "We have the whip-hand and you can use it. I pray that you will for our own national safety. . . . Keep your eyes peeled on the British at the same time that you are seeing them through."[27] Interestingly, Elliott later mentored Henry Kissinger in the art of statecraft.[28]

In contrast to Elliott's growing interest in mobilizing the American public to recycle, other key officials in the Roosevelt administration remained opposed. When Senator James J. Davis of Pennsylvania wrote to the president in April 1941 to suggest a nationwide salvage drive, FDR solicited the advice of Leon Henderson, who led the Office of Price Administration. Henderson told him that "any effort to stimulate a popular collection of scrap would probably do more harm than good." Henderson feared that it might disrupt the flow of materials to the scrap trade and thus drive up prices. Echoing the position that Elliott had held earlier, Henderson concluded, "The time may come when we would want such a popular scrap collecting campaign. For the present at least I feel that efforts in this direction should be discouraged."[29]

Although British leaders expressed profound gratitude for Lend-Lease aid, they worried about what the Americans would demand in return for it. Speaking to his own advisers soon after the law passed, the British ambassador to the United States, Lord Halifax, asserted, "The President is so obviously badly informed as to our real financial position" that the British had an urgent need to head off a possible announcement on his part about repayment arrangements. Halifax's staff urged him to make it clear to Roosevelt "that it will be sheer impossibility" for the British to repay the Americans after the war with "money or goods; . . . our hope is that the United States will regard the fact that we continue fighting as the true consideration for the benefits we receive under the Act."[30] Later that

[27] William Yandell Elliott to Averell Harriman, 7 Mar. 1941 (copy), Harriman Papers, Box 158, Folder 5.

[28] Jeremi Suri, *Henry Kissinger and the American Century* (Cambridge: Belknap Press of Harvard University Press, 2007), 112.

[29] Leon Henderson to FDR, 3 May 1941, President's Official File, FDR Library, Salvage 1941–43 folder.

[30] Aide-mémoire by Edward Frederick Lindley Wood, 1st Earl of Halifax, 31 Mar. 1941, Halifax Papers, Borthwick Institute for Archives, University of York, A4.410.4.2.4. I am grateful to Lord Halifax for granting me permission to use the Halifax Papers.

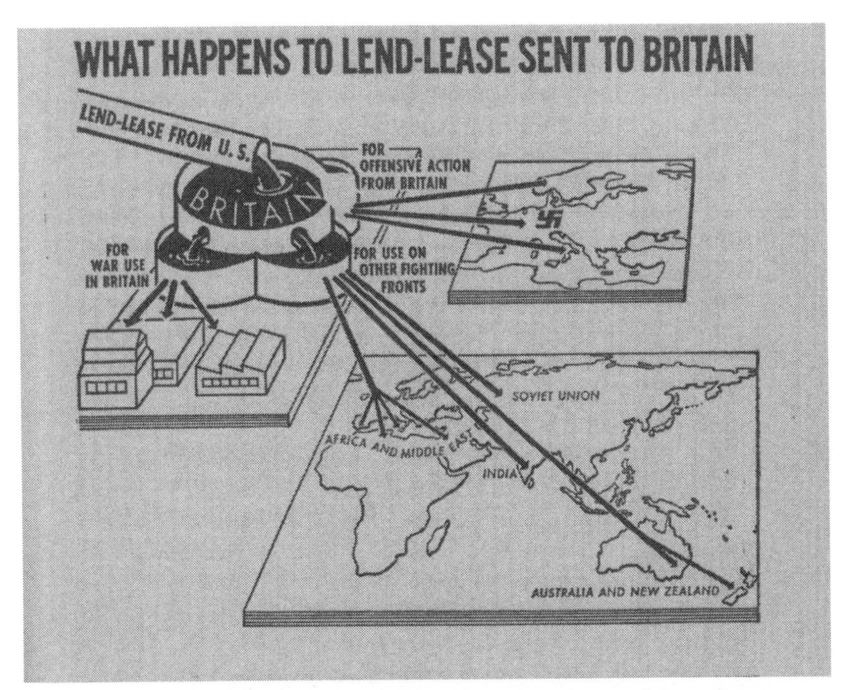

FIGURE 10. "What Happens to Lend-Lease Sent to Britain." In an effort to maintain public support for Lend-Lease, U.S. officials sought to demonstrate that Britain was making good use of every dollar it received.
Source: Edward R. Stettinius, Jr., *Lend-Lease: Weapon for Victory* (New York: Macmillan, 1944), 243.

day, Halifax was more diplomatic with FDR, telling him, "We have been fighting the costliest war in history for eighteen months, half the time alone. Enormous restrictions have been placed on civilian consumption of every kind and taxation and borrowing at home have been pushed to lengths which no one would have thought possible before war started. . . . Anyone is blind who does not realise that great national impoverishment later must follow."[31]

King George VI followed these developments carefully, and his correspondence provides a fascinating glimpse into the complexities of Anglo-American relations during this period. Writing to Halifax less than a month after Lend-Lease became law, the king praised Roosevelt, noting,

[31] Lord Halifax to FDR, 31 Mar. 1941 (copy), Halifax Papers, Borthwick Institute for Archives, University of York, A4.410.4.2.1–2.

"He is certainly the right man for the job, + he is getting a move on with it." The king also put to rest any lingering doubts about Harriman, whom he described after their first meeting as "very much on our side. . . . We are assembling a good team of people on both sides of the Atlantic, which should make for a proper understanding of our views." Doing so was essential, he suggested, because the terms of the Lend-Lease Act, enacted less than a month earlier, put the United States in a very powerful position:

> I did not feel too happy about the Lease of the Bases as the Americans wanted too much written + laid down. Everything was done in their interests, no give + take in certain circumstances. But no doubt when they get to know us better, as they will, we shall be able to settle things with them. You must be very thankful the Lease + Lend Bill was passed at last, with few amendments. I do hope that the Americans will not try + bleed us white over the dollar asset question. As it is they are collecting the remaining gold in the world, which is of no use to them, and they cannot wish to make us bankrupt. At least I hope they do not want to.[32]

Harriman quickly got to work in London, assisted by Robert Meiklejohn, with whom he had worked at the Office of Production Management. Describing the headquarters of the Harriman Mission, as it soon became known, Meiklejohn noted, "We are comfortably installed now in our own offices in the same building as the American Embassy, with direct connection to the Embassy itself. Mr. Harriman achieves a somewhat Mussolini-like effect – not at all to his liking – by reason of his office being a very large room that used to be the living room of a rather elegant flat. I am located in what I believe was the dining room, with a kitchen connecting. It is rather odd for an office, but still very comfortable and works out quite satisfactorily."[33]

One of the first things that Harriman did upon his arrival in London was to seek as much information as possible about the economy. On 9 April 1941 he contacted Thomas H. Brand, an official with the British government's North American Supply Committee. "It would be most helpful to me," Harriman wrote, "in understanding the basis of the requests which your Government may from time to time put

[32] George VI to Lord Halifax, 4 Apr. 1941, Halifax Papers, Borthwick Institute for Archives, University of York, A2.278.26.4. Extracts from this letter are reproduced with the kind permission of Her Majesty Queen Elizabeth II.

[33] Robert Meiklejohn to Knight Woolley, 21 May 1941 (copy), Harriman Papers, Box 159, Folder 5.

forward to mine for supplies, if your Government will arrange to make members of my staff acquainted with . . . statistical information on the subject."[34]

Not surprisingly, Harriman's request received prompt attention. The morning after Harriman sent it, the economist Sir Walter Layton, then serving in the Ministry of Supply, phoned the cabinet secretary, Sir Edward Bridges, to discuss how to proceed. Layton recommended giving Harriman full access to the British government's own compendium of highly classified data, but Bridges raised a series of objections: "Before doing so, we should obtain the consent of the Departments concerned, namely the Ministry of Supply and the Ministry of Food." Bridges also wanted to "make sure that Professor Lindemann, who is the Prime Minister's adviser on this question of what should be given to the U.S. authorities, agrees with this course." Even if these hurdles were cleared, Bridges made it clear that he had no desire to be completely candid with the Americans. "I do not think that the line should be to give the whole of the Tables to Mr Harriman. The proper course, in my view, would be that someone should show him these Tables and ask him which of them he wants."[35]

While other British officials were debating how forthcoming they ought to be with the U.S. government, Layton quietly went ahead with his work on a summary of Britain's raw materials situation for Harriman. According to a draft version of this document, the British government had imposed stringent control over consumption, and supplies of many resources were "being eked out by careful salvage. In the case of newsprint, for example . . . consumption is now on a basis of less than one fourth of normal and will only be maintained at a level which limits standard newspapers to 4 pages a day, by drawing on stocks and by using a high percentage of waste paper in English mills." The United Kingdom faced "the double task of feeding Britain and developing our maximum war effort on about half the weight of imports of normal times. . . . This is a stiff proposition for a country which ordinarily imports some 60 per cent of its food supplies, whose home ore production yields not more than 30 per cent of its annual steel production, . . . which

[34] Averell Harriman to Thomas H. Brand, 9 Apr. 1941 (copy), NA, CAB 139/88. The British spelling of "programme" likely crept in when this letter was retyped in London.

[35] Edward Bridges, "Sir Walter Layton's Memorandum to Mr Harriman," 10 Apr. 1941, NA, CAB 139/88.

FIGURE 11. Beaverbrook visits Washington. In August 1941, immediately after the Atlantic Conference between FDR and Churchill in Newfoundland, Britain's minister of supply, Lord Beaverbrook (shown here with his hands together), met in Washington with Edward R. Stettinius Jr. (priorities director of the Office of Production Management, whom Roosevelt would soon appoint to lead the Office of Lend-Lease Administration) and other American officials. Averill Harriman, America's top Lend-Lease official in London, stands at the right in this photo. It was during this visit that the Americans informed Beaverbrook of their decision to halt exports of scrap iron and steel to the United Kingdom. Beaverbrook promptly ordered the requisition of iron gates and railings across Britain.
Source: Farm Security Administration/Office of War Information. Courtesy of the Library of Congress, Prints and Photographs Division, LC-USE6-D-000964.

produces virtually no non-ferrous metals and little timber, and whose only abundant raw material is coal." The report concluded with a stark assessment: "We cannot both feed our people on a standard that will maintain stamina and health and also work up our war effort to its planned peak and keep it there on the present level of imports. But we *can* contrive to perform this double task for several months by killing off a substantial part of the animal population and by running down stocks of food and material to a level which leaves little or no margin for setbacks or exceptional losses."[36]

By 1941 the United Kingdom had practically run out of money with which to pay for imports, despite raising taxes and borrowing heavily from its colonies. Some British officials argued that Britain should be completely frank about this crisis, but others disagreed. Lindemann, for instance, told Churchill that the memo Layton had prepared "is to my mind painted in exaggeratedly sombre colours, but presumably he thought this necessary in order to impress on Harriman the urgent need for more American ships."[37] In the months that followed, Harriman used the whip that Roosevelt and Elliott had handed him to demand more and more details about the British economy and its military strategy. Writing to Roosevelt a month after he left Washington, Harriman informed him that British officials "who are in a position to know fully appreciate there is little hope except with American aid."[38] The need for U.S. support soon prompted British officials to swallow their pride and share information with the United States that they would have preferred to keep to themselves.

American public opinion was a topic of supreme importance to British officials in the spring of 1941. Four days after Harriman arrived in London, the director of broadcasting relations in the Ministry of Information reported that an "urgent need" existed "for a change in tone of our publicity in and towards America." Asserting that "the 'Britain Can Take It' theme has been overplayed," he maintained that the time had come for Britain to adopt a more aggressive stance. "Everybody likes to back a winning horse and the more we can convince America that we are bound to win the more enthusiastic and whole-hearted support we shall receive." To accomplish this end, he suggested two new slogans, neither of which was exactly inspiring: "Britain Will Win," and "Britain Can Produce and Deliver the Goods." The memo went on to suggest ideas for films, such as *Night Flight to Berlin*, which was to depict "in a realistic way (even if it has to be faked) the drama of a British bombing raid on a German city."[39]

[36] "Note on Imports, Shipping and Supplies Prepared for Mr. Harriman," n.d. (ca. Apr. 1941), 5, 8, NA, BT 87/40.

[37] Frederick Lindemann to Winston Churchill, 13 Dec. 1940 (copy), Cherwell Papers, Nuffield College Library, Oxford, F.124/2; Frederick Lindemann to Winston Churchill, 22 Apr. 1941 (copy), Cherwell Papers, Nuffield College Library, Oxford, F.222/26.

[38] Averell Harriman to FDR, 10 Apr. 1941 (copy), Harriman Papers, Box 159, Folder 1.

[39] Notes re. Policy Committee, 19 Mar. 1941, NA, INF 1/196.

Despite being responsible for distributing millions of dollars each week in aid, Harriman often felt left in the dark about developments in Washington. In a letter in which he asked the president for better communication if he was to do his job properly, Harriman complained, "The mail pouch is intolerably slow in delivery – three to four weeks." Although he received telegraph messages from Washington in "answer to specific questions," his only other "source of information is entirely from the British Ministries. I send a man daily to each Ministry with which I am dealing to look over the cable interchange between them and Washington."[40] Six weeks later, after the communication situation had failed to improve, Harriman asked the U.S. ambassador in London, John Winant, for help. "I wish Harry [Hopkins] could understand how much more valuable we could be to him if he would have someone write me a weekly cable, followed by a letter in more detail, advising me of what is going on in Lend-Lease. . . . Our information now comes from the British, as I see in each of the Ministries their communications in both directions with Washington. Naturally I would prefer to see Washington through American rather than British eyes."[41]

On 11 April 1941, Harriman traveled by train with Churchill and Winant to Bristol, where they arrived "a few hours after the Huns had finished a 6-hour raid – one of the worst in a badly bombed city. . . . The P[rime] M[inister] inspected the damaged area, walking among the people and visiting the reception centers. A grim determined people but he was met everywhere with cheers and smiles. I talked to many of them."[42] In a letter to FDR several weeks later, Harriman wrote, "The Prime Minister continues to take me with him on his frequent trips to the devastated cities. He thinks it of value to have an American around for the morale of the people."[43]

Writing at the end of May to his friend William Jeffers, president of the Union Pacific Railroad, Harriman boasted, "The work here is fascinating. I am in constant touch with the Prime Minister and all of the other Service and Supply Ministers. My activities cover a wide field; shipping, naval, aircraft and ordnance, supply, and also food and other civilian problems. I am accepted practically as a member of the Cabinet,

[40] Averell Harriman to FDR, 10 Apr. 1941 (copy), Harriman Papers, Box 159, Folder 1.

[41] Averell Harriman to John Winant, 27 May 1941 (copy), Harriman Papers, Box 159, Folder 6.

[42] Averell Harriman to FDR, 11 Apr. 1941 (draft copy), Harriman Papers, Box 159, Folder 1.

[43] Averell Harriman to FDR, 7 May 1941 (copy), Harriman Papers, Box 159, Folder 5.

and am given all information. It gives me an opportunity to follow the war from day to day."[44]

BARBAROSSA

On 22 June 1941, Hitler launched Operation Barbarossa, the largest invasion in world history. Along a front thousands of miles wide, four million Axis soldiers crossed into the Soviet Union in an attack that took Stalin completely by surprise. So confident had the Soviet leader been in his treaty of nonaggression with Nazi Germany that just a few hours before the invasion began, he ordered the execution of a German deserter who had warned Soviet forces of the imminent attack. Fortunately for the deserter, the invasion commenced before the death sentence could be carried out.[45]

Three days after Barbarossa began, J. C. Dawes, the deputy controller of salvage, sent a memorandum to every local authority in Britain. Without mentioning the USSR by name, Dawes wrote, "The war situation calls imperatively for an intensification of salvage work and the fullest co-operation of the [Town] Councils and the public." He went on to announce that the government would soon "make it an offence to destroy or throw out or put into refuse receptacles any waste materials."[46] That same day, salvage officials in the district that encompassed the Birmingham region called on the Ministry of Supply to postpone making recycling compulsory for individuals until local councils had time to make arrangements to deal with the range of materials that citizens would contribute. Many communities, they maintained, faced severe shortages of labor and transport and simply could not process any more salvage.[47]

Many Britons greatly admired the people of the Soviet Union as they fought to repell the German invasion of their country. Even Churchill, despite his strong aversion to Communism and distrust of Stalin, famously remarked that "if Hitler invaded hell, I would at least make a

[44] Averell Harriman to William Jeffers, 30 May 1941 (copy), Harriman Papers, Box 159, Folder 6.

[45] Norman Davies, *No Simple Victory: World War II in Europe, 1939–1945* (New York: Penguin, 2007), 160.

[46] Ministry of Supply, Salvage Circular 55 (25 June 1941), 1, East Sussex Record Office, Brighton, C/C/55/132.

[47] Ministry of Supply, Salvage Department, Area No. 3, Minutes of a meeting in Birmingham, 25 June 1941, 4, Shakespeare Centre Library and Archive, Stratford-upon-Avon, BRR 55/14/31/4/1.

favourable reference to the Devil in the House of Commons."[48] Britain simply had no chance of defeating Germany without the help of the USSR. Lindemann, who recently had received the title of Baron Cherwell, took a similar view. Writing to Churchill in September 1941 he noted that the Red Army was inflicting "probably 25,000 casualties on Germany every day, the equivalent of 2 months of Bomber Command's full effort. To keep Russia fighting is a most economical way of killing Germans." Lindemann also recognized the enormous sacrifices that the Soviet people were making: "In every two days the Russians suffer more losses than our army has incurred in two years of the war."[49]

In October 1941 Lord Beaverbrook, whom Churchill had appointed minister of supply four months earlier, called on the people of Britain to redouble their salvage as a way to help the Soviet Union. A short time later an advertisement in the *Times* reinforced this message. "You owe Russia every scrap of paper in your home," declared its headline. The accompanying illustration, created by the popular artist Gilbert Rumbold, reflected the widely shared assumption that "the enemy of my enemy is my friend." The image depicted a Red Army soldier pressing against a phalanx of German tanks, artillery, and bayonets, all aimed at the British Isles. Disregarding the pleas of historians and archivists, the advertisement carried an insistent message: "Lord Beaverbrook has got to have *all your stored up paper*. All the old forgotten books – all the old treasured programmes – the useless receipts – sentimental letters – historic newspapers. . . . Turn them out in cold blood. Be sentimental about Russia – about Britain – about Freedom – but forget sentiment about yourself."[50] Such messages were not limited to advertisements. On New Year's Day 1942 the editors of the *Times* urged their readers to salvage Christmas wrapping paper so it could be recycled into weapons.[51]

As localities across the United Kingdom responded to the call to collect salvage in support of the USSR, many developed their own ways to encourage contributions. In November 1941, the lord mayor of

48 Fraser J. Harbutt, *The Iron Curtain: Churchill, America, and the Origins of the Cold War* (Oxford: Oxford University Press, 1986), 132.

49 Frederick Lindemann to Winston Churchill, 11 Sept. 1941 (copy), Cherwell Papers, Nuffield College Library, Oxford, F.224/65–6. Extracts from the Cherwell Papers are reproduced with the kind permission of the Cabinet Office Historical and Records Section.

50 "You Owe Russia Every Scrap of Paper in Your Home," *Times* (London), 7 Nov. 1941, 7. Emphasis in original.

51 "Waste Paper Contest: Opening of New Year Campaign," *Times* (London), 1 Jan. 1942, 8.

Manchester sponsored an event called Anglo-Soviet Friendship Week. The festivities included a parade in which twenty-two trucks full of recycled materials were adorned with placards that proclaimed "Bury Hitler with Salvage," and "Save Russia, Save Britain, Save Your Salvage."[52] In Westminster, an anonymous philanthropist promised to give £10 to charity for every ton of paper people donated within a single week. To inform residents of this fact without consuming paper in the process, officials used a loudspeaker van.[53]

In addition to promoting increased salvage, the Ministry of Supply unleashed a raft of regulations that restricted the use of paper to a greater extent than ever before. Reporting on this development, the *Times* told its readers, "There must be no new Christmas cards. Parcels must be unlabelled. Posters must be reduced to half their size, and to the previously existing ban on blowing one's nose on paper is added a ban on cleaning one's face with paper." The new order also prohibited the publication of any magazine or newspaper that had not existed prior to August 1940.[54]

In the autumn of 1941, Beaverbrook reorganized the Salvage Department and renamed it the Directorate of Salvage and Recovery. The new terminology reflected a growing emphasis on extracting resources not only from the waste stream but also from objects that people would never have discarded in peacetime. To oversee its work, Beaverbrook appointed a new leader, Mr. G. B. Hutchings, and created a new administrative title for him: director of salvage and recovery. Hutchings, whose title soon grew to *principal* director of salvage and recovery, remained in charge of the directorate until 1945, when he took a job in another government department.[55] The arrival of Hutchings amounted to a demotion for Harold Judd and J. C. Dawes, the men who had created the Salvage Department and led it for the first two years of the war. Judd became responsible for salvage from industry, and Dawes became director of salvage at the local authority level. Both had to report to Hutchings.[56]

[52] "Making Manchester Salvage Conscious," *Waste Trade World and the Iron and Steel Scrap Review* (hereafter *WTW*), 22 Nov. 1941, 10–11.

[53] "Waste Paper Drive in Westminster," *Times* (London), 16 Dec. 1941, 2.

[54] "Saving Paper," *Times* (London), 12 Nov. 1941, 5. For a detailed summary of wartime paper control measures, see Ministry of Supply, R.M.D. (Regional) Memorandum No. 31, Part 7, 23 Feb. 1942, NA, SUPP 14/631.

[55] "News in Brief," *Times* (London), 2 Oct. 1941, 4; "Domestic Salvage Campaign," *Times* (London), 11 Nov. 1942; "News in Brief," *Times* (London), 22 Feb. 1945, 8.

[56] "New Controller of Salvage (Local Authorities)," *Public Cleansing*, Jan. 1942, 126; Ministry of Supply, "Memorandum on British Methods of Raw Material Conservation," Sept. 1943, 22, NA, AVIA 46/490.

Beaverbrook also created an advisory group called the Salvage and Recovery Board, chaired by the economic expert Sir Vyvyan Board (who after the war became known for his advocacy of electrified harpoons as a "humane" method of killing whales).[57] Other members of the board included Stanley Bell, managing director of Associated Newspapers; Harold Judd; Robert Morrison, MP, chairman of the Waste Food Board; and A. H. Read, an official with the British Iron and Steel Corporation.[58] Writing to his colleague Wyndham Portal, who served jointly with him as parliamentary secretary to the Ministry of Supply, Macmillan suggested in October 1941 that Megan Lloyd George and her colleagues on the Women MPs' Salvage Advisory Committee might serve as "a very useful buffer to us" against criticism if the government decided to mandate recycling for the general population. Macmillan also told Portal that

> the women M.P.s are very much concerned about the formation of the Salvage Board and think it means that their position is threatened. . . .
>
> If we ever contemplate compulsory ideas, these women M.P.s may be a very useful buffer to us.
>
> They held a meeting yesterday which, I gather, was very stormy. Lady Davidson, Miss Ward and Mrs. Rathbone are (I am told) particularly excited. Personally I do not much like M.P.s and certainly not women M.P.s, But there it is.
>
> I have reason to believe that this "storm in the tea-cup" (or rather "refuse bin") can be calmed by inviting Miss Lloyd George to take a seat on the Salvage Board. I cannot myself see any reason why this would be undesirable.[59]

A short time later, Megan Lloyd George accepted an invitation to join the board.[60] There she faced continued sexism. As she noted in her own diary, one board member responded to her first appearance there by exclaiming, "Good Lord! This is a business-meeting!" Despite this hostile welcome, the man restrained himself and the rest of the "meeting passed

[57] "A Salvage Board," *Times* (London), 2 Oct. 1941, 4; "Sir Archibald Vyvyan Board," *Times* (London), 12 Jan. 1973, 15; Patrick Geddes, "Sir Vyvyan Board," *Times* (London), 24 Jan. 1973, 18.

[58] "Salvage and Recovery Board," *Times* (London), 10 Oct. 1941, 5.

[59] Harold Macmillan to Wyndham Portal, 24 Oct. 1941 (copy), Harold Macmillan Papers, Department of Special Collections and Western Manuscripts, Bodleian Libraries, University of Oxford, c. 273, fol. 164. Extracts from the Harold Macmillan Archive are reproduced with the kind permission of the Trustees of the HM Book Trust. On Portal, see J. V. Sheffield, "Portal, Wyndham Raymond, Viscount Portal (1885–1949)," rev. Robert Brown, in *Oxford Dictionary of National Biography* (Oxford: Oxford University Press, 2004).

[60] "Miss Lloyd George on Salvage Board," *Times* (London), 7 Nov. 1941, 2.

off remarkably well. I'm just not so sure about their salvage methods however!"[61]

STETTINIUS

When he signed Lend-Lease into law in March 1941, Roosevelt put his close friend and political adviser Harry Hopkins in charge of managing its operation. Yet Hopkins soon realized that the job of administering America's foreign-aid program was more than he could manage in light of his fragile health and his ongoing work as FDR's closest adviser. Hopkins recommended his friend Edward R. Stettinius Jr. for the job, and the president agreed.[62] Roosevelt had appointed him previously to lead the Industrial Materials Department of the National Defense Advisory Commission and later as priorities director of the Office of Production Management.[63] Stettinius was by all accounts an extraordinary man. Two years earlier, at the age of thirty-seven, he had become chairman of U.S. Steel, one of the largest corporations in the world. Stettinius possessed not only intelligence and ambition, but also an ebullient personality. As a profile in *Time* magazine put it,

> He calls people by their first names, which he always remembers. He chews gum, smokes cigarets [*sic*], smiles often. He has an almost pastorlike skill at presiding over meetings. He has a knack for getting people to agree. He leans back, crosses his legs, talks informally. A caller at his office is greeted like a long-lost brother; Ed sits down facing him, slapping his big hands down on both knees, leaning forward, all interest. He has presence. He is tall, handsome and prematurely white-haired. The color of his hair, and his quick rise to position, long ago gave him a nickname, not always spoken in jest: the White-Haired Boy.[64]

Upon hearing that Stettinius's salary would be $10,000 a year (quite a large figure at the time), presidential speechwriter Samuel Rosenman

[61] Megan Lloyd George, Diary, 1941, 11 Nov. 1941, National Library of Wales, Aberystwyth, MS. 23138B. Reproduced by kind permission of Llyfrgell Genedlaethol Cymru/The National Library of Wales.

[62] Sherwood, *Roosevelt and Hopkins*, 376–7. In contrast to Hopkins and Harriman, historians have paid relatively little attention to the career of Edward R. Stettinius Jr. (1900–49). For details of his life, see Thomas M. Campbell and George C. Herring, introduction to *The Diaries of Edward R. Stettinius, Jr., 1943–1946* (New York: New Viewpoints, 1975).

[63] Report of Industrial Materials Department . . . June 1 to Aug. 31, 1940, 3, Stettinius Papers, Box 89; "President Delegates Power to End White House Delay," *New York Times*, 17 Sept. 1941, 1.

[64] "Mr. Secretary Stettinius," *Time*, 11 Dec. 1944, 20.

FIGURE 12. Stettinius confers with Roosevelt. Edward R. Stettinius Jr., whose father had played a key role in supplying the Allies with American arms during the First World War, played a similar role during the Second World War as head of the Office of Lend-Lease Administration. Stettinius subsequently served as secretary of state and as the United States' first ambassador to the United Nations.
Source: Farm Security Administration/Office of War Information. Courtesy of the Library of Congress, Prints and Photographs Division, LC-USE6-D-007317.

commented to Grace Tully, FDR's private secretary, that this would mean a massive pay cut for Stettinius. "$10,000," he quipped, with little exaggeration, "is about his daily income."[65]

Despite the financial sacrifice that it entailed, Stettinius jumped at the opportunity to devote himself entirely to government service. In late October, the organization that he now led changed its name from the Division of Defense Aid Reports to the Office of Lend-Lease Administration (OLLA).[66] One of the first items to cross Stettinius's desk in his new

[65] Samuel Rosenman to Grace Tully, n.d. [ca. 15 Sept. 1941], President's Official File, FDR Library, Office of Lend-Lease Administration 1941 folder.
[66] "Roosevelt Signs New Leasing Bill," *New York Times*, 29 Oct. 1941, 2.

job was a letter from Sir Clive Baillieu, director of the British Purchasing Commission, based in Washington. "I hope," he wrote, "you will feel that the measures which have been taken to meet the position in the United Kingdom, ensures [*sic*] best possible use being made of available supplies."[67] This was a highly charged issue, for then as now, many Americans were highly skeptical about foreign aid programs. Just a month before Stettinius took the helm, Major General J. H. Burns, a top official in the Division of Defense Aid Reports, informed the White House that there was a "widespread campaign to discredit the Lend-Lease program by implying that the United States is being made a 'sucker' by Great Britain. . . . Among the rumors that are being spread are such things as that we are buying meals and wine for the British Purchasing Commission . . . and that they don't really need what they are getting."[68] Administration officials publicly dismissed such reports as anti-British propaganda, but they privately made it clear to the British that it was in their interest to demonstrate that they were making judicious use of all resources at their disposal.

For this reason, Stettinius was concerned when he received reports in fall 1941 that British steel production was operating far below capacity. Stettinius raised the matter with Morris Wilson, who had recently succeeded Arthur Purvis as chairman of the British Supply Council in North America after the latter perished in a plane crash. Wilson demanded an explanation from his staff. They soon confirmed that the U.S. critique was accurate: Britain's factories had the potential to produce 14.5 million gross tons of steel each year, but they were making only twelve million.[69] Six months later George B. Waterhouse, professor of metallurgy at the Massachusetts Institute of Technology, provided an even bleaker assessment of steel production in the United Kingdom. In a confidential report that Averell Harriman had asked Waterhouse to prepare for the War Production Board, the professor noted that in contrast to the United States, where steel production was operating at full capacity, British steel mills were "only running at about 72 to 73% of capacity." Waterhouse attributed the problem to shortages of iron ore and scrap, the two primary ingredients in steel. In 1937, Britain had imported seven million tons of high-grade iron ore from Sweden, Spain, Morocco, and elsewhere, but imports

67 Clive Baillieu to Edward R. Stettinius Jr., 5 Sept. 1941, NARA, Record Group 169, PI 29 9, Box 162.

68 Summary of communication from J. H. Burns to Stephen Early, 25 Aug. 1941, President's Official File, FDR Library, British Supply Council in North America 1941–43 folder.

69 I. F. L. Elliot to Morris Wilson, 15 Dec. 1941, NARA, Record Group 169, PI 29 9, Box 167.

had fallen sharply since the outbreak of war. In 1940, Britain imported only 4.7 million tons of iron ore, and in 1941 the figure fell to a mere 2.3 million tons. Through a great deal of effort, British miners had increased their output of iron ore from 13.1 million tons in 1937 to 18.1 million tons four years later. Despite these impressive gains, domestic mines proved incapable of making up for the decline in imports of iron ore, in part because the iron content of British ore was far lower than that of the foreign ore it replaced.[70]

Although Britain imported nearly a million tons of iron and steel scrap from the United States in 1940, American scrap shipments to Britain declined sharply in 1941.[71] That April the British learned that "only very limited tonnages can be made available."[72] The problem had two causes, according to a British official who was closely involved in the matter: "Scrap is in short supply in the United States owing partly to the growing domestic demand and partly to the fixing by the United States Government of the price of scrap at too low a level which has caused dealers to 'hold off.'"[73] The United States, which had long been a major exporter of scrap iron and steel, was soon importing it from over a dozen other countries.[74] With demand outstripping supply, the price of scrap on the global commodities market rose sharply. Britain managed to import only half a million tons of iron and steel scrap in 1941. For the remainder of the war, the United Kingdom obtained no significant quantities of iron or steel scrap from beyond the British Empire.[75] To make up for the decline in scrap imports, British officials redoubled their efforts to obtain scrap from Great Britain and the empire – and to make tough decisions about how to use it.[76]

Steel played a far more extensive role in the Second World War than in any previous war, and even the most advanced economies struggled to produce enough to meet the demands of their militaries. Edwin Barringer, the president of a trade association that represented the American scrap iron and steel industry, declared in 1942, "This is a steel war. Notwithstanding the growing importance of the lighter metals, steel is still

[70] George B. Waterhouse to Averell Harriman, 26 June 1942, Stettinius Papers, Box 158.

[71] *Institute of Scrap Iron and Steel Yearbook, 1940*, 33–4; *Institute of Scrap Iron and Steel Yearbook, 1942*, 19, 40.

[72] Telegram from Elliot to Read, sent 8 Apr. 1941, NA, CAB 115/123.

[73] Note to Cyril Hurcomb, 28 July 1941, NA, MT 59/865.

[74] Note to Arthur B. Van Buskirk, 28 May 1943, NARA, Record Group 169, PI 29 8, Box 134.

[75] Peter Howlett, *Fighting with Figures: A Statistical Digest of the Second World War* (London: Central Statistical Office, 1995), 9.10.

[76] Central Office for North American Supplies, "Empire (ex Canada) Requirements and Supplies of Raw Materials, July 1941–Dec. 1943," n.d. [Sept. 1941], NA, BT 87/52.

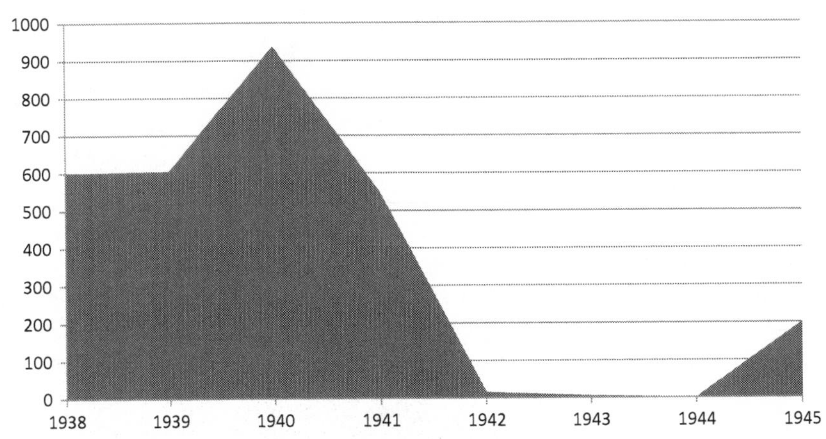

FIGURE 13. Imports of iron and steel scrap to the United Kingdom (thousand tons), 1938–45.
Source: Peter Howlett, *Fighting with Figures: A Statistical Digest of the Second World War* (London: Central Statistical Office, 1995), 9.10.

either the material for or the means of manufacturing the sinews of war – the tank, the warship, the jeep, the shell, the gun, the machine tools by which planes and other materiel are produced." Barringer added that technological changes meant that each soldier now needed "an average of 4,900 pounds of steel in the form of carried or supporting equipment," far more than had been required in the Great War.[77] By 1943, armaments production consumed 85 percent of Britain's steel output.[78]

In December 1941, a member of the British Purchasing Commission stationed in Washington expressed the hope that Stettinius would understand that the British were "getting every pound [of scrap] which can be collected, including 'blitz' scrap and recovery of scrap from wrecks at sea, under conditions of the greatest hazard." The author urged Stettinius to inform his countrymen "that the British really are doing all that is possible, but that they are severely restricted on the raw material side, particularly in respect of importation of iron ore. Also that they could operate with steel furnaces nearer capacity, if America could supply them with pig iron and scrap, but that there is a shortage in America also, and the British have agreed not to press for further supplies."[79]

77 *Institute of Scrap Iron and Steel Yearbook*, 1942, 27.
78 Joel Hurstfield, *The Control of Raw Materials* (London: HMSO, 1953), 105.
79 I. F. L. Elliot to Morris Wilson, 15 Dec. 1941, NARA, Record Group 169, PI 29 9, Box 167.

Because of the shortage of steel, the Home Office and Ministry of Home Security saw their allocations of steel cut by 50 percent for the first quarter of 1942. "I am afraid," noted one official in a secret report to the War Cabinet's Production Executive in December 1941, "that with the present shortage a postponement of the programme of Indoor Shelters is inevitable."[80] Although the recycling of iron and steel made it possible for Britain to virtually eliminate its reliance on imported scrap by 1942, such a policy could not continue indefinitely. As one American expert told Harriman in June of that year, Britain's "scrap supply is being kept up by drawing from stocks that cannot be replenished." He estimated that officials were melting down 6,000 tons of blitz scrap and 10,000 tons of railing every week.[81]

PEARL HARBOR

In November 1941, the U.S. Office of Production Management established a Bureau of Industrial Conservation, later called the Salvage Division.[82] Its initial plans were quite modest, but this soon changed when, on 7 December 1941, Japan attacked the U.S. Pacific fleet at Pearl Harbor, and Germany joined it in declaring war on the United States. In sharp contrast to the largely dismissive attitude toward recycling that had prevailed within the U.S. government during previous two years, America's direct involvement in the war prompted officials to embrace salvage wholeheartedly. As they did so, they borrowed heavily on British organizational models and promotional methods.

Just two days after Pearl Harbor, Donald Nelson, head of the Office of Production Management (soon to be the War Production Board) informed Roosevelt that he was organizing a campaign to collect scrap for the war effort. "The President encouraged this," noted Stettinius, "and thought it ought to be on a very bold front."[83] A short time later,

80 J. J. Llewellin, "Steel: Proposed Allocation for the First Quarter, 1942; Supplementary Report by the Chairman of the Materials Committee," 19 Dec. 1941, appendix, Beaverbrook Papers, Parliamentary Archives, London, BBK/D/111.

81 George B. Waterhouse to Averell Harriman, 26 June 1942, Stettinius Papers, Box 158.

82 "Additional Report of the Special Committee Investigating the National Defense Program . . . Interim Report on Steel,"4 Feb. 1943, 78th Congress, 1st session, Report No. 10, Part 3, 13.

83 Edward R. Stettinius Jr., notes of a meeting at the White House on 9 Dec. 1941, Stettinius Papers, Box 137. In 1942 Nelson became chair of the War Production Board after Roosevelt established it in January 1942 to replace the Office of Production Management and the Supply Priorities and Allocations Board.

the U.S. government organized a "national scrap harvest," which brought forth two million tons of ferrous metal from American farms, and called on retailers and wholesalers to "make a clean sweep-up of critically needed scrap materials."[84] As in Britain, children played an important part in the U.S. strategy. To encourage them to help, the Conservation Division of the War Production Board produced a booklet for the nation's schools entitled *Get in the Scrap*. Inside its front cover was a copy of a letter from President Roosevelt, in which he asserted that "the boys and girls of America can perform a great patriotic service for their country by helping our National Salvage effort." This campaign, the brochure explained, provided "the greatest opportunity for every civilian to back up our soldiers on the firing line." Just as the British had called on children to join a salvage army, so too did the Americans. Each classroom teacher would have the rank of major, and children, if they displayed the requisite effort and results, could be promoted from the beginning rank of private to that of corporal, sergeant, or lieutenant. Another way that the United States mimicked British precedent was by encouraging the public to imagine specific weapons that could be made from salvaged materials. *Get in the Scrap*, for instance, claimed that a metal pail contained enough steel to produce three bayonets, and that an old lawn mower could be made into six three-inch shells.[85]

On the night of 12 December, less than a week after the attack on Pearl Harbor, a secret train pulled away from platform six of London's Euston Station. After traveling through the night, the train arrived at Greenock, Scotland, where its passengers boarded a battleship, HMS *Duke of York*, for a journey to the United States. In addition to Churchill, the group included Lord Beaverbrook (minister of supply), Sir Dudley Pound (first sea lord), Field Marshal Sir John Dill (head of the army), Major-General G. H. Macready (second-in-command of the army), and Sir Charles Portal (chief of the air staff). Accompanying this high-powered delegation were several Americans, including the ambassador, John Winant, and FDR's Lend-Lease "expediter" to Britain, Averell Harriman.[86]

84 "Strategic Materials," *A Week of the War*, 15 Aug. 1942, 4, NARA, Record Group 169, PI 29 83, Box 685.

85 War Production Board, *Get in the Scrap* (Washington: U.S. Government Printing Office, 1942), frontispiece, 5, 7, 11.

86 L. C. Hollis, "Operation 'Arcadia'. Administrative Arrangements," 11 Dec. 1941, Beaverbrook Papers, Parliamentary Archives, London, BBK/D/98; Brian Lavery, *Churchill Goes to War: Winston's Wartime Journeys* (London: Naval Institute Press, 2007), 78–9.

While crossing the Atlantic, Beaverbrook wrote a memo that outlined the additional assistance that he hoped the United States would supply to Britain. The document dealt with both manufactured goods and raw materials. Among the latter, steel was at the top of the list. "For the completion of our 1942 war programme we need 12 million tons of finished steel. To achieve this we must get from the US about 4 million tons of steel, 300,000 tons of pig iron and 500,000 tons of scrap. This assumes we are able to maintain our imports of iron ore at the present level (2½ million tons) and also our shipments of pig iron and steel from the Empire."[87] In actuality, the United Kingdom would manage to import only 1.9 million tons of ore per year in 1942 and 1943. The decline was even more drastic in the case of iron and steel scrap. British imports of scrap from all sources beyond its empire, which rose sharply during the first year of the war to 900,000 tons, had plummeted since then. From 1942 through the end of the war, the only scrap that entered the British Isles came from its own colonies.[88]

After a circuitous and stormy journey, the *Duke of York* anchored in Chesapeake Bay on the afternoon of 22 December.[89] Just two days later, a member of Beaverbrook's staff who had traveled with him to Washington observed, "The objections of U.S. authorities and of [the] U.S. steel industry to giving us scrap or pig iron have strengthened since last August." As a result, advised the raw materials expert Charles Morris, "It would not be wise to press for scrap and pig."[90]

On 27 December Beaverbrook invited Stettinius to his hotel, where he tried from the start to impress and intimidate his American counterpart. Stettinius's notes of the meeting are quite revealing. "When I got to the Mayflower, he had a palatial suite and was surrounded by stacks of papers and three secretaries. . . . He conversed with me and dictated at the same time." As was his habit, Stettinius recorded the quality of the dining experience, noting that "we were beautifully served with huge slabs of roast beef and mashed potatoes by waiters from the battleship."

[87] "Memorandum by Lord Beaverbrook for Prime Minister and Chiefs of Staff," 19 Dec. 1941, 7, Beaverbrook Papers, Parliamentary Archives, London, BBK/D/111.

[88] Peter Howlett, *Fighting with Figures: A Statistical Digest of the Second World War* (London: Central Statistical Office, 1995), 213.

[89] Lavery, *Churchill Goes to War*, 85.

[90] "Scrap and Pig-Iron – 1942," 24 Dec. 1941, Beaverbrook Papers, Parliamentary Archives, London, BBK/D/111. This version of the document contains three short paragraphs and is unsigned; an adjacent copy in the same file consists of six numbered paragraphs and is signed C. R. Morris.

Despite the fact that Beaverbrook "set right out attacking me" for not doing more to help Britain, Stettinius appreciated his forthright manner and described him as "very friendly and very cooperative. We had a frank, genuine talk about the whole thing."[91]

At the same time that Churchill and Beaverbrook were in Washington, an important meeting took place in London to discuss the future of salvage. The chair of the meeting was Lord Reith, formerly the director-general of the BBC and now the minister of works. Reith's goal for the meeting was to coordinate the efforts of the various government departments that were participating in the search for iron and steel scrap. Reith began the meeting by reading from two telegrams, labeled "most secret," which Beaverbrook had sent from Washington. The first of these asserted,

1. We shall not get any more scrap in 1942 from the United States and you must develop your programme on that assumption.
2. The Americans collect razor blades. The yield may not be great but the psychological effect is important and we should do the same.
3. The Americans are pulping their records of the last war. We should try to follow their example.
4. The whole subject of scrap collection is more vigorously handled in the United States than in Great Britain. There is a much better show of publicity here, apart from the waste paper publicity in the United Kingdom. The newspapers should help in a general scrap campaign in which razor blades might be the first item.[92]

In his second telegram, Beaverbrook suggested that to make up for the scrap iron and steel that the United States was no longer exporting, the United Kingdom would have to collect 40,000 tons of scrap a week for the foreseeable future. "We must gather in all scrap, however preciously prized, including old guns on village greens. . . . In particular, we must search out and make use of structural steel, of which there must be quantities."[93]

Reith, who rarely saw eye to eye with Beaverbrook, told those present that he wanted to "concentrate on big stuff. Razor blades and village dumps

[91] Edward R. Stettinius Jr., notes of a meeting with Lord Beaverbrook on 27 Dec. 1941, Stettinius Papers, Box 137.

[92] Lord Beaverbrook to Wyndham Portal, 31 Dec. 1941, Beaverbrook Papers, Parliamentary Archives, London, BBK/D/111; Ian McIntyre, "Reith, John Charles Walsham, first Baron Reith (1889–1971)," in *Oxford Dictionary of National Biography* (Oxford: Oxford University Press, 2004).

[93] Lord Beaverbrook to Wyndham Portal, n.d. [likely Dec. 1941], Beaverbrook Papers, Parliamentary Archives, London, BBK/D/111.

might come later." He also announced that he planned to requisition metal "from all derelict buildings, factories, railways, redundant machinery, and general scrap, lying about in factories, coal mines, shipyards, builders' yards etc." This was a huge step for the government to take. Crucially, the minutes of the meeting recorded that the participants promised to implement the new regulation "even where constitutionally it might not apply."[94]

When the War Cabinet discussed Reith's requisition plans a short time later, some argued that it would be unfair to pay owners only scrap value for spare parts or machinery that would cost far more than that to replace. Despite such reservations, requisition of industrial equipment began in many parts of the country in April 1942.[95] By that September, the government required occupiers of all premises throughout Britain with three or more tons of scrap metal to report their holdings or face heavy fines. Government agents had the right to enter places to seize undeclared metal, although owners had the right to appeal the removal of machinery and equipment.[96]

Just as he had with Hopkins and Harriman during their visits to Britain, Churchill sought during his time in the United States to cultivate a warm relationship with Stettinius. In January 1942, at FDR's request, Stettinius lent his vacation house in Pompano Beach, Florida, to Churchill, who spent six days there relaxing in complete secrecy. On his return to Washington, Churchill spoke privately with the Lend-Lease director. In his record of this meeting, Stettinius noted that Churchill made a point of saying that he wished "to tell me something . . . that occurred during the last war in relationship with my father that had meant much to him from the standpoint of confidence." The prime minister explained that there had been "an important contract between the U.S. Government and the British government and in the final accounting there was a dispute over the settlement. He stated that he had worked many weeks with his accountants and lawyers preparing a 73-page digest of the case. . . . He stated that my father opened the report, glanced at the opening page and replied that he had complete confidence and trust in Churchill as a man and

[94] The agenda for this 5 Jan. 1942 meeting and notes by several of those who were present can be found in NA, HLG 102/94 and AVIA 15/2573.

[95] War Cabinet, Home Policy Committee (Legislative Section), 20 Jan. 1942, NA, HLG 102/94; Statutory Rules and Orders 1942, No. 761, Emergency Powers (Defence): Scrap Metal (Particulars), 23 Apr. 1942, NA, HLG 102/94.

[96] "Scrap Metal to Be Registered: New Order in Force To-Day," *Times* (London), 21 Sept. 1942, 2; "Drive for Scrap Metal: Stocks of Three Tons or More Must Be Disclosed," *Public Cleansing and Salvage*, Oct. 1942, 64.

wished to accept his accounting as final. The Prime Minister said this had touched him deeply and that at that time, when his standing in England was not very great, to have an American display this confidence and trust in him was something he could never forget. He became emotional and his eyes filled with tears when he described this incident."[97]

Both British and U.S. officials were extremely concerned about American attitudes toward the United Kingdom. In May 1941, just two months after the Lend-Lease began, Churchill's adviser Professor Lindemann warned him of American "suspicions, only smouldering now but always ready to be re-kindled, that the infirmities of mind and will displayed during the appeasement period are too deep-seated to be cured."[98] Stettinius later detected a similar trend, warning Harriman that there existed "quite a feeling against the British developing over the country. This is, of course, attributed to Axis-inspired propaganda. The Administration is aware of the situation and is doing everything possible to allay this feeling."[99]

One of the important ways that American officials did this was to press the British to recycle greater quantities not only of civilian articles but also of battle scrap. A good start had already been made. In September 1940, the Ministry of Aircraft Production appointed an official to coordinate the recycling of crashed airplanes, both British and German.[100] Applauding this development, an editorial in the *Times* noted that Germany had "made a most useful contribution to our supplies of scrap metal. The graveyards of destroyed German aeroplanes which have been brought down on our soil have proved and are proving mines of aluminium and of other metals, and there is a particular satisfaction in that the enemy is helping to provide the means which will eventually procure his defeat."[101] In addition to recycling metal from downed planes, the British government raised from the seabed a number of battleships that Germany had been forced to scuttle after the First World War under the disarmament clauses of the Treaty of Versailles. Churchill's science adviser Frederick Lindemann

[97] Edward R. Stettinius Jr., "Record of Meetings with the President and Prime Minister Churchill on Saturday, January 3rd and Tuesday, January 13th, 1942," 4, Stettinius Papers, Box 125.

[98] Frederick Lindemann to Winston Churchill, 7 May 1941 (copy), Cherwell Papers, Nuffield College Library, Oxford, F.134/10. Extracts from the Cherwell Papers are reproduced with the kind permission of the Cabinet Office Historical and Records Section.

[99] Edward R. Stettinius Jr. to Averell Harriman, 4 Mar. 1942, NARA, Record Group 169, PI 29 83, Box 685.

[100] "In Brief," *Manchester Guardian*, 24 Sept. 1940, 6.

[101] "Feeding the Factories," *Times* (London), 23 Sept. 1940, 5.

informed him in January 1941 that during the previous six months the salvage of ships had made a greater contribution by weight to the war effort than had the building of new ships. He also noted that between September 1939 and December 1940, salvaged ships accounted for 43 percent of the tonnage added to the British merchant fleet.[102]

Cartridge cases constituted another military source of recyclable material. Whenever a fighter plane fired a machine gun, its spent cartridges fell to the ground. For each minute of fire from all eight guns, 9,600 cartridge cases, weighing 240 pounds, were ejected. Bombers, with more space, retained their spent cartridges, which made it possible for them to be recycled.[103] Most cartridge cases did not even need to be melted down to be reused. Workers in a dozen locations in Britain who in peacetime had manufactured cookware or maintained railway equipment refurbished an average of over 100,000 brass cartridge cases each week. The used cartridges went through a multistep process in which they were cleaned, heated, pressed, and refinished.

In July 1941, while on a five-week trip to the Middle East, the American official Averell Harriman sent a secure cable to Churchill from the British embassy in Cairo to offer his "impressions and reactions" on the British war effort there. Assessing the subject of military supplies, he complained, "There has been too much living from hand to mouth. . . . There are no adequate arrangements for salvage of tanks – a deficiency which added materially to the losses in the recent engagement."[104] Churchill evidently took Harriman's critique seriously. Between October 1941 and September 1942, the British government salvaged an enormous quantity of military scrap, including 3,450 tons of shell cases from the North African campaign. To reduce the risk of explosion, used cartridges were "muffled" in a furnace that was at least 550 degrees Celsius, a temperature sufficient to detonate "duds" and to evaporate some of the mercury that remained from the firing charge (thereby increasing the purity of the brass).[105] Airplanes, ships, and shell casings were not the only military items to be

[102] Salvage of Merchant Ships, n.d., Cherwell Papers, Nuffield College Library, Oxford, F.124/22; Frederick Lindemann to Winston Churchill, 6 Jan. 1941 (copy), Cherwell Papers, Nuffield College Library, Oxford, F.220/111; memorandum, 20 Mar. 1941, NA, BT 87/40.

[103] "Scrap from the Sky," *WTW*, 3 Aug. 1940, 4.

[104] Averell Harriman to Winston Churchill, 1 July 1941 (copy), Harriman Papers, Box 160, Folder 1.

[105] "Copper: Interim Report No. 3," 27 Jan. 1943, 27 Jan. 1943, Appendix 4, NARA, Record Group 169, PI 29 9, Box 167.

salvaged. The military also collected enormous quantities of animal bones and fats, broken-down equipment, and empty containers of all sorts.[106]

Researchers devoted considerable efforts to finding ways to make munitions from recycled materials. One of the most plentiful items in Britain's arsenal was the .303 Mark VII bullet. Its primary ingredient was lead, supplemented with small amounts of other metals for hardness. A million rounds of this bullet required 9.25 tons of lead. British military experts anticipated the need for 720 million rounds of this bullet in 1942, the production of which would consume 555 tons of lead each month. Prior to the Second World War, arms makers in Britain used virgin lead for the manufacture of bullets. Motivated by a desire to reduce imports, the government explored the feasibility of making bullet cores from recycled lead. In addition to saving shipping space and reducing the risk of interrupted supplies, the use of recycled lead was expected to save money.[107] Lead was far from the only ingredient in weapons that could be supplied from recycled materials. In 1942, the Ministry of Supply commissioned a private firm to study methods of manufacturing gun metals and phosphor bronzes not only from lower grades of tin but also from "a higher proportion of scrap metal."[108]

The U.S. and British governments forged a strong and close relationship during the Second World War, but they did not always agree about how Lend-Lease funds should be spent. As a case in point, in April 1942 U.S. officials rejected a British request for 240 million sanitary napkins. In light of the severe shortage of paper in Britain and the importance of minimizing women's absences from war work, the provision of menstrual pads would no doubt have been of great assistance to Britain. For reasons of prudery, however, the Lend-Lease policy committee turned down the request. This decision, explained Thomas McCabe, who was serving as acting administrator during a period in which Stettinius was recuperating from ill health, "is not based on the opinion that it may not be necessary or to the advantage of the direct war effort, but it is the type of transaction which, because of the public reaction here, should be handled, we believe, by other means than Lend-Lease."[109]

[106] "Salvage Work in the Army," *Times* (London), 12 Jan. 1943, 2.

[107] Memorandum, 13 Nov. 1941, NA, AVIA 22/2349.

[108] Ministry of Supply, Contract with Tin Research Institute, 3 Oct. 1942, NA, AVIA 22/2357.

[109] Thomas McCabe to Averell Harriman, 15 Apr. 1942, NARA, Record Group 169, PI 29 83, Box 685.

6

Waste Becomes a Crime

> Modern war absorbs the whole of the economic and financial resources of the nation, non-combatants as well as the men actually put into uniform. It is not a business to be carried on by a section of the population, even by a large section; it requires the efforts of all.
>
> – Geoffrey Crowther, 1940[1]

Although the British government long resisted compelling individuals to recycle their rubbish, this changed on 9 March 1942 when the Salvage of Waste Materials (No. 2) Order took effect. This regulation made it a crime to burn, throw away, or mix with refuse any paper or cardboard.[2] The *Times* praised the order. Asserting that mandatory recycling was long overdue, an editorial claimed that "British citizens at every stage of this war have been ready for a stricter discipline than Governments have been ready to apply."[3] Shortly after the new regulation took effect, the newspaper commented that "litter was a peace-time nuisance; in war-time it is a criminal waste of material that is vitally important in the manufacture of munitions."[4]

The trade journal *Public Cleansing*, welcoming the fact that the government had finally made it illegal "to treat paper or cardboard as rubbish," denounced as "salvage saboteurs" the "hundreds of thousands

[1] Geoffrey Crowther, *Ways and Means of War* (Oxford: Clarendon Press, 1940), 2.

[2] The Salvage of Waste Materials (No. 2) Order was issued as S.R. & O. 1942 No. 336. See Great Britain, *Statutory Rules and Orders 1942*, vol. 2 (London: HMSO, 1943), 1991. For a discussion of this order, see "Salvage of Paper," *Municipal Review* 13 (Mar. 1942): 48.

[3] "Waste of Paper an Offence: Stringent New Order," *Times*, 5 Mar. 1942, 2; "Precious Waste," *Times* (London), 5 Mar. 1942, 5.

[4] "Punishing Litter Offenders: Names to Be Taken," *Times* (London), 10 Mar. 1942, 2.

of people in this country who are helping the enemy every day. They are the . . . householders who continue to throw into the dust-bin valuable salvage material that will help to make munitions, armaments and equipment for the Fighting Forces, and stuff that will feed animals and help to grow precious crops."[5] Charles Peat, parliamentary secretary to the Ministry of Supply, went even further, declaring that "the man or woman who destroyed a piece of paper was sabotaging the war effort. He or she was as bad as any fifth columnist."[6]

J. C. Dawes, who oversaw Britain's municipal salvage efforts, announced in spring 1942 that since the start of the war local authorities had collected two and a quarter million tons of recyclable materials. This total included 639,000 tons of paper, 527,000 tons of iron and steel, 35,000 tons of rags, 19,000 tons of bones, and 271,000 tons of kitchen waste. In addition, Britain's households had recycled 886 tons of aluminum, 2,235 tons of brass, 2,382 tons of copper, and 5,340 tons of lead.[7] A few days later, the *Times* noted that some of the materials that had been reclaimed for the war effort came from surprising sources. A single railway horse depot, it claimed, had yielded a hundred tons of horseshoes.[8]

To help craft its efforts to promote salvage, the Ministry of Information conducted an opinion survey of the public in the early months of 1942. The interviewees consisted of 3,078 housewives from nine large towns, seven small towns, and twelve rural areas. Only 60 percent of those surveyed thought that officials were making good use of salvage.[9] One month later Sir Andrew Duncan, who recently had returned as head of the Ministry of Supply, asked the nation's top recycling official, G. B. Hutchings, to undertake a comprehensive review with the goal of increasing the amount of material that people contributed to salvage collections. This assignment came at a time of widespread frustration and confusion about salvage among members of the public, merchants, and elected officials.

5 "Editorial," *Public Cleansing*, Apr. 1942, 208.

6 "'Private Scrap' in London: Host at a Salvage Exhibition," *Times* (London), 18 Dec. 1942, 6.

7 J. C. Dawes, "Making Use of Waste Products," *Journal of the Royal Society of Arts* 90 (15 May 1942): 388–408, esp. 393 and 397.

8 "Salvage on the Railways: Reclamation of Plant and Paper," *Times* (London), 20 Apr. 1942, 2.

9 Ministry of Information, *Wartime Social Survey: Salvage* ([London]: n.p., 1942), 1, 4, copy in NA, RG 23/9B.

Less than a month later, Hutchings presented a forty-one page report. In it, he pointed out that the country's salvage efforts suffered from a lack of clear coordination. Many government departments had a role in salvage activities, and none seemed willing to take direction from any other. In Hutchings's view, "The principal difficulties which have arisen . . . have concerned the activities of the Ministry of Supply and the Ministry of Works and Buildings." Duncan's own agency, the Ministry of Supply, was home to no fewer than four entities that dealt with salvage: the Directorate of Salvage and Recovery (in charge of household waste), the salvage branch of the Raw Materials Department (responsible for "commodity salvage" of by-products or cuttings from manufacturing processes), the Timber Control, and the Iron and Steel Control (which handled "the heavier types of iron and steel scrap").[10]

Because the Ministry of Supply lacked the equipment and workforce needed to collect heavy scrap, it relied upon the Ministry of Works to do this task. As Hutchings explained, "This includes the recovery of steel from buildings destroyed or damaged by enemy action (outside the London region), the removal of railings and tram rails, and the recovery of iron and steel from obsolete and redundant buildings, plant and machinery."[11]

Hutchings argued that the turf battle between these two ministries, which had grown especially fierce during the time that Beaverbrook had led the Ministry of Supply, caused confusion and inefficiency. The general public, local government bodies, and voluntary groups were perplexed about the respective areas of responsibility of each agency. Instead of coordinating their efforts, the two ministries appeared to compete against each other. As an example, Hutchings noted that every summer since the beginning of the war, salvage officials in Whitehall had organized county-level drives designed to generate fresh enthusiasm for recycling. Seemingly unaware of this practice, the Ministry of Works had conducted "overlapping" campaigns of its own.[12]

STETTINIUS VISITS BRITAIN

On 14 July 1942, Stettinius left the United States on a fact-finding mission to Britain. His goal was to ascertain British needs and to assess what the

[10] G. B. Hutchings, Report on Salvage, 18 May 1942, 1, 2, 6, NA, AVIA 11/22.
[11] G. B. Hutchings, Report on Salvage, 18 May 1942, 6–7, NA, AVIA 11/22.
[12] G. B. Hutchings, Report on Salvage, 18 May 1942, 9, NA, AVIA 11/22.

TABLE 3. *Works Departments*

Office of Works (until Oct. 1940)

First Commissioner of Works and Public Buildings

Herwald Ramsbotham, MP (June 1939–Apr. 1940)
Earl de la Warr (Apr. 1940–May 1940)
Lord Tryon (May 1940–Oct. 1940)

Secretary

Sir Patrick Duff

Ministry of Works and Buildings (Oct. 1940–June 1942)

Minister

John Reith, first Baron Reith (Oct. 1940–Feb. 1942)
Wyndham Portal, Baron Portal (appointed Feb. 1942)

Parliamentary Secretary

George Hicks, MP (appointed Nov. 1940)

Secretary

Sir Patrick Duff (until May 1941)
William Leitch (acting secretary May–Nov. 1941)
Sir Geoffrey Whiskard (appointed May 1941, but absent until Nov. 1941)

Director General, Works and Buildings

Hugh Beaver (appointed Apr. 1941)

Ministry of Works and Planning (June 1942–Feb. 1943)

Minister

Wyndham Portal, Baron Portal

Joint Parliamentary Secretaries

George Hicks, MP (Works)
H. G. Strauss (Planning)

Secretary

Sir Geoffrey Whiskard

Director General, Works and Buildings

Hugh Beaver

Ministry of Works (from Feb. 1943)

Minister

Wyndham Portal, Baron Portal (until Nov. 1944)
Duncan Sandys, MP (Nov. 1944–July 1945)

Parliamentary Secretary

George Hicks, MP

Secretary

Sir (Frederick) Percival Robinson

Director General

Hugh Beaver (became Sir Hugh Beaver in June 1943)

Sources: John D. Cantwell, *The Second World War: A Guide to Documents in the Public Record Office*, 3d. ed. (Kew: Public Record Office, 1998), 238–9; *Times* (London), various dates.

British were doing to support the war effort. Shortly before Stettinius departed for the United Kingdom, one of his advisers, Arthur B. Van Buskirk, suggested that Harriman had become overly sympathetic to the British position. He predicted that once in the United Kingdom, Stettinius would see that "Mr. Harriman favors giving the British tremendous freedom in determining civilian needs, but many of us feel that perhaps we owe the duty to do a more thorough job ourselves. . . . I believe Congress will take this view." The memo went on to argue that an urgent need existed to dispel anti-British sentiment in the United States.[13]

After arriving in Britain, Stettinius passed through Bristol, the city that Harriman had visited with Churchill just hours after a major attack in April 1941. As over a year had passed since the last heavy raids on the city, Stettinius assumed that all salvageable materials would be cleared away by the time he arrived. Instead, "upon . . . seeing blitz scrap uncollected all over the city, I obtained a most unfavorable first reaction to the British scrap collection effort."[14] Attempting to regain control of the message two days later, a civil servant in the Board of Trade urged his colleagues to show the Lend-Lease director "where the shoe pinches + where it doesn't. It is thought that Mr. Stettinius has not had this matter properly put to him + that now is an opportunity of enlightening him to our advantage."[15] Despite British officials' hopes that they could manipulate Stettinius, the American would demonstrate that he was even more adept than they at wielding influence.

Upon reaching London, Stettinius met with the prime minister and Mrs. Churchill; their son-in-law Duncan Sandys, MP; and the American diplomat William Bullitt, who had earlier served as U.S. ambassador to the Soviet Union and to France. Stettinius's notes of the meeting are revealing: "Great atmosphere and tradition – 10 Downing Street. . . . Met in small library. Presented ham. Mrs. Churchill and Prime Minister delighted. Sherry, and lunch served promptly in small dining room with tiny table for five. Butler, two maids in white gloves. Silver immaculate. . . . Stayed with Prime Minister from 1:30 to 3:45, last hour alone. Most frank and confidential. Talked of Father again. . . . Prime Minister showed strain since January when I saw him last at the White House – eyes."[16]

[13] Arthur B. Van Buskirk to Edward R. Stettinius Jr., 11 July 1942, Stettinius Papers, Box 149.

[14] Edward R. Stettinius Jr., "Lend-Lease, Reciprocal Lend-Lease and the British War Effort," [1942], 5, Stettinius Papers, Box 149.

[15] J. J. Wills, handwritten note, 18 July 1942, NA, BT 11/2078.

[16] Edward R. Stettinius Jr., notes, 16 July 1942, Stettinius Papers, Box 125.

Officials throughout the British government worked hard to ensure that Stettinius would view their country in a positive light. Yet their efforts to give him a grand welcome seemed to contradict the impression they wished to convey: that they faced significant shortages and required large amounts of American aid. On 18 July 1942, the British government hosted an opulent dinner in Stettinius's honor. The occasion appalled John Maud, a top official in the Ministry of Food. In a report to Lord Woolton, the minister of food, Maud noted, "We ate soup, lobsters, chicken, asparagus, and raspberries and cream with five different wines, in a private room at Claridge's." In Maud's view, the "non-austerity" of the meal "was a great mistake, as Stettinius, [Tony] Biddle (late U.S. Ambassador [to the Polish government in exile] in Paris) and two or three American officials from the Harriman Office were among the guests. . . . Both Biddle and Stettinius left a large part of the poultry course uneaten on their plates."[17]

The next day, Stettinius set out on foot to explore the war's effect on London: "Cellars of bombed houses converted into water tanks. . . . Churches badly gone. Park grounds unkept [*sic*]." He also "made many observations on iron fences, steel scrap, brass signs."[18] When he incorporated these notes into the draft of his book about Lend-Lease, Stettinius (or more probably, his ghostwriter, Louis Hector) wrote, "I had noticed many ornamental brass signs, lead gutters, and other sources of metal scrap which were not being exploited. After talking the situation over with the British officials, I had to tell them that I was not satisfied that there was any good reason for failing to collect this scrap also. There were shortages of all these metals in the United States, and we could not be expected to supply them under Lend-Lease if they were not making the most of their own sources of supply."[19]

Among the many officials whom Stettinius met in London were Charles Morris and Sir William Palmer. According to his own detailed

[17] John Maud to Lord Woolton, 18 July 1942, Papers of Frederick James Marquis, 1st Earl of Woolton, Department of Special Collections and Western Manuscripts, Bodleian Libraries, University of Oxford, MS. Woolton 12. Not to be confused with Sir John Maude, who was secretary to the Ministry of Health. See Robert Armstrong, "Maud, John Primatt Redcliffe, Baron Redcliffe-Maud (1906–1982)," in *Oxford Dictionary of National Biography* (Oxford: Oxford University Press, 2004).

[18] Edward R. Stettinius Jr., notes, 19 July 1942, Stettinius Papers, Box 148.

[19] Edward R. Stettinius Jr., draft of Lend-Lease book, chpt. XXII, 15, Stettinius Papers, Box 213. For evidence of Hector's role in the drafting of this book, see Stettinius Papers, Box 199.

notes of the meeting, Stettinius asked them to keep him "informed about salvage operations, specifically with respect to steel scrap, copper and brass, rubber and ships." When Stettinius asked what they were doing to increase their utilization of scrap, they told him that

> the salvage of steel from . . . damaged buildings, railings, etc. is being pressed. They began with the places near the steel mills and have worked out to more distant points. Mr. Morris pointed out that the collection of scrap from damaged buildings required a vast amount of labor and was a risky job and Sir William stated that the amount recovered was less than would seem likely from inspection of the buildings themselves. . . . Substantial amounts of copper scrap have been collected but there is no electrolytic refinery in the U.K. so that the lower grade scrap such as door-knobs, brass name-plates, etc. cannot be made useful in the U.K. Arrangements are being made, however, to ship some to the U.S. to be refined there.[20]

A report originally designated "U.S. Secret – British Most Secret," which contains notes of Stettinius's meeting with the enormously influential British economist and Treasury official John Maynard Keynes on 21 July 1942, mentions that Keynes told him that the United Kingdom was borrowing £5 million a week from India and £2 million a week from Egypt.[21] These were vast sums, and even a small fraction of this money would have done much to help the millions who died of famine in Bengal the following year.[22] In a letter that he sent Stettinius shortly after this meeting, Keynes invited the American to "take supper with my wife and myself in our kitchen at Gordon Square."[23]

In addition to telling Stettinius about their efforts to recycle various materials, British officials emphasized that civilian consumption had been reduced to a bare minimum. Compared to 1938 levels, British factories were producing only 30 percent as many razor blades, only 20 percent as many "invalid carriages" (wheelchairs), and only a third as many bicycles. Production of civilian automobiles had ceased entirely, along with many other articles, including bathroom fixtures, musical instruments, metal furniture, toys, spring mattresses, metal bedsteads, and sports equipment.[24]

[20] Edward R. Stettinius Jr., notes, 20 July 1942, 2, Stettinius Papers, Box 148. Britain did in fact possess some refineries that could recover nonferrous scrap, but these factories lacked the capacity needed to process large quantities.

[21] Edward R. Stettinius Jr., notes, 21 July 1942, Stettinius Papers, Box 148.

[22] For details of this famine, see Lizzie Collingham, *The Taste of War: World War II and the Battle for Food* (New York: Penguin, 2012), 141–54.

[23] John Maynard Keynes to Edward R. Stettinius Jr., 24 July 1942, Stettinius Papers, Box 150.

[24] "The Present Basis of Steel Consumption in the United Kingdom: Appendix A," 22 July 1942, 2, Stettinius Papers, Box 148.

On 24 July Stettinius's deputy, Thomas McCabe, cabled the Lend-Lease director to inform him that a crisis had developed during his absence from the United States. "Labor leaders and newspapers are beginning to be aroused about shortage of steel in many plants and this whole matter is becoming a serious problem." Because of this shortage, construction workers had been forced to suspend progress on a massive new shipyard in New Orleans. McCabe consequently urged his boss to "look carefully into all angles of steel requirements in the U.K. before you return."[25]

The very next day Stettinius told Andrew McCosh, an official in the Iron and Steel Control, that he "look[ed] forward to hearing" about the progress that the British made in "the clearing of 'Blitz' scrap in Bristol."[26] McCosh responded by noting that because the railway network in that area was extremely burdened, "as much scrap as possible has to be moved by sea." This was a problem, however, because there was also a shortage of ships. While acknowledging that "it is unfortunate that there should be any accumulation of scrap anywhere," he argued that only a tiny percentage of blitz scrap remained to be cleared.[27] Two days later, in a letter to the U.S. embassy official William Bullitt, Stettinius informed him that "a complete study of the whole scrap situation is being made."[28]

Stettinius maintained a hectic schedule during his 1942 visit to London. Perhaps his busiest day was Thursday, 6 August. That morning he met with Keynes; Baron Catto, financial adviser to the Treasury; and Brendan Bracken, minister of information – all before attending the weekly staff meeting in Harriman's office at the American embassy. Stettinius then had lunch at the Savoy with *New York Times* reporter Raymond Daniell and three dozen others, after which he met separately with Churchill and with Charles Portal, who led the Royal Air Force. At 8 P.M. he attended another sumptious dinner at Claridge's, hosted this time by the Coventry industrialist Sir William Rootes.[29]

Before he left Britain, Stettinius asked for a detailed inventory of the efforts that the government had taken to collect recyclable materials for

[25] Thomas McCabe to Edward R. Stettinius Jr., 24 July 1942, Stettinius Papers, Box 135.
[26] Edward R. Stettinius Jr. to A. K. McCosh, 25 July 1942 (copy), Stettinius Papers, Box 148.
[27] A. K. McCosh to Edward R. Stettinius Jr., 29 July 1942, Stettinius Papers, Box 148.
[28] Edward R. Stettinius Jr. to William C. Bullitt, 31 July 1942 (copy), Stettinius Papers, Box 148.
[29] Appointment cards for 6 Aug. 1942, Stettinius Papers, Box 148.

the war effort. In response, McCosh informed him that as of 4 July 1942, the British had collected 409,243 tons of blitz scrap, almost 300,000 tons of railings, 485,000 tons of old or obsolete equipment and structures, and 90,000 tons of tram rails.[30] Yet these numbers were little more than guesses. Only six days later, in fact, McCosh sent a revised estimate to Stettinius in which he claimed that workers had salvaged a total of 533,000 tons of blitz scrap by 11 July.[31] As a report issued after the war explained, "In the earlier days . . . the work was handled by scores of completely inexperienced men engaged by the Ministry of Works. Thousands of tons of material were collected and despatched without any record being taken. In many instances records were incomplete and woefully inaccurate, and weighings were not taken. The situation was in truth chaotic."[32]

In his 1944 book, *Lend-Lease: Weapon for Victory*, Stettinius noted that "several things about the British iron and steel picture" had disturbed him during his visit to Britain two years earlier.

> Near the American Embassy, I saw day after day the huge girders of some buildings which had been wrecked in the air raids many months before. What bothered me most, however, was to see the high iron fence around the park in Grosvenor Square every time I walked out of the United States Embassy. It seemed to me that the British were not making the most of their scrap collections. . . . I told them that we could not continue to supply them under Lend-Lease at the present rate unless they were collecting all the scrap iron and steel they possibly could at home. . . . It was good news to get the message a few months later that "the railings in Grosvenor Square are down at last."[33]

In addition to cutting down vast quantities of gates and railings, the British government pulled up thousands of tons of tram rails. Because they were often embedded in roadways – and covered with asphalt after being abandoned – tram rails were a much more expensive source of metal than were railings. Despite the high costs involved, officials believed that the country could not afford to ignore them.[34] R. M. Hunter, an official with the Ministry of Works, noted in August 1942 that the

[30] A. K. McCosh to Edward R. Stettinius Jr., 23 July 1942 (copy), Stettinius Papers, Box 148.

[31] A. K. McCosh to Edward R. Stettinius Jr., 29 July 1942, Stettinius Papers, Box 148.

[32] British Iron and Steel Corporation (Scrap), Report to Directors, Board Meeting, 28 Nov. 1945, 1, NA, AO 30/32. Emphasis in original.

[33] Edward R. Stettinius Jr., *Lend-Lease: Weapon for Victory* (New York: Macmillan, 1944), 251–2.

[34] Ministry of Supply, Memorandum No. 315/44, 28 Oct. 1944, NA, POWE 5/114.

government hoped to obtain at least 750 tons of tram rails every week. "This scrap which is of known specification and quality is required for bomb manufacture and I understand there is nothing to take its place." Hunter explained that the British Iron and Steel Corporation, which had been set up by the government, paid £8 per ton for tram rails. Because so much work was required to remove these rails and repair the roads from which they were taken, the total cost of recovering a ton of tram rails frequently reached £40 a ton. The Ministry of Works, Hunter noted, had been "inundated with requests to proceed with the lifting of tram rails and is subject to considerable pressure on account of the steps which have been taken to recover more readily obtainable scrap, e.g. railings, gates, etc. whilst ignoring this source of supply."[35]

Bath spent £2,500 removing tram rails for the war effort, but it obtained only £220 in exchange for those rails – a transaction that the local press labeled a "bad bargain."[36] In Bristol, the city engineer informed municipal leaders that the costs of removing all of its old tram rails and restoring the torn up roadways would cost £50,000 more than it received for selling the rails.[37] Facing pressure to help with these costs, the Iron and Steel Control and the London Passenger Transport Board provided subsidies to local councils. Even with this assistance, the borough councils of Hackney and Islington had to borrow £34,967 and £14,474, respectively, to cover the shortfall.[38]

On his return to the United States, Stettinius praised the "full and detailed information" that British officials had shared with him. "In a comparatively short time," he recalled, "it was possible to obtain a first-hand understanding of the British war effort and the relation to it of lend-lease operations." Stettinius added that he was confident that "the British nation and its people have gone 'all out' in their war program."[39] "One of the most important questions" involving Lend-Lease, he noted, "was the manner in which war control . . . has been established over all important raw materials. With the demand on our own economy due to our own war program, it is of primary importance that the best possible use be made of the available resources."[40] Stettinius went on to discuss

35 R. M. Hunter to H. E. C. Gatliff, 6 Aug. 1942, NA, T 162/748.

36 "£220 Worth of Tram Rails Costs Bath £2500!" *Bath Weekly Chronicle and Herald*, 24 May 1941, 15.

37 "Bristol Tram Rails Salvage Costs," *Western Daily Press and Bristol Mirror*, 26 Aug. 1943, 3.

38 London County Council, *Minutes of Proceedings, 1943–1945*, 9 Nov. 1943, 286.

39 Stettinius, "Lend-Lease," 1.

40 Stettinius, "Lend-Lease," 4.

the salvage question, particularly with respect to steel scrap, rubber, copper and brass. Salvage operations are, in some cases, ahead of those in the United States. For instance, the British are using a large amount of reclaimed rubber in military tires. Rubber collections have been such that they have exceeded the U.K. capacity to handle reclaim material, and many thousands of tons of reclaimed rubber have been sent to the United States for reclamation.[41]

Stettinius also complimented the British for doing much to make up for the cessation of steel scrap deliveries from the United States. He estimated that the military accounted for three-quarters of all steel consumption in the United Kingdom and asserted that "practically all of the remaining 25% is for indirect but essential war purposes, such as maintaining operations in mines, railways, and public utilities, all of which have been cut far below their normal rate of use for essential maintenance and repair purposes."[42] Stettinius expressed great admiration for the way in which ordinary Britons had risen to the challenges of war, noting,

The attitude of complete trust and tremendous appreciation on the part of the British toward the United States was most heartening and reassuring. I was convinced the British people as well as their high government representatives are thoroughly alive to the tasks of the future. They have already willingly undergone incredible privations and an almost complete "leveling" of living standards. The British have accomplished a truly amazing transposition of their economy to an all out war effort, making full and effective use of all available materials with considerable ingenuity and at the cost of great sacrifice. [43]

At the same time that he expressed his admiration for the British people and their leaders, Stettinius emphasized one point above all else: aid to the United Kingdom was not charity but was an investment that directly promoted the national security of the United States. "Without lend-lease aid the British economy would collapse and with such a collapse the burden of total war against the Axis aggressors might well be more than our own economy and manpower could cope with." Employing the same metaphor for Lend-Lease that he would later use in the title of his wartime book on the subject, he referred to it as "a weapon of prime importance to this country and our Allies."[44]

In a report that the Ministry of Works shared with Stettinius in 1942, its officials noted that collections of blitz scrap were dependent "on the amount of bomb damage occurring in any period." Because most blitz scrap had been cleared, they predicted that railings would constitute the

[41] Stettinius, "Lend-Lease," 5.
[42] Stettinius, "Lend-Lease," 6–7.
[43] Stettinius, "Lend-Lease," 20.
[44] Stettinius, "Lend-Lease," 21.

major source of scrap iron and steel for the duration of the war. According to their estimates, another quarter million tons of railings might be harvested.[45] Yet just a week after he returned to the United States, Stettinius informed a delegation of six British officials that "from his observations of the scrap situation in England, . . . the British would be able to obtain sufficient scrap iron from the areas in Britain which had been devastated by bombing." This source of scrap was not so plentiful as to allow Britain to export scrap iron to the United States, he explained, but it ought to cover British needs for the time being. Stettinius declared that for this reason, the United States "should not . . . have to send any scrap to the U.K. at least for some months."[46]

A short time later, the MIT professor George Waterhouse informed Stettinius that the British were going to "spectacular" lengths to scrap the railings of "parks, homes, churches, etc., which is bringing the war very close to home to the people of Great Britain." Although not all of the material was of high quality, Waterhouse wrote, a substantial proportion of it was suitable for making steel. The rest was used in blast furnaces to make pig iron. Waterhouse observed that "many women are used to supervise the removal of railings and to organize the collection of scrap in villages and what are called 'county sweeps.'" In his view, the British public generally supported the government's policies on scrap, but he noted that "there has been a good deal of criticism in regard to the railing scrap due to the inevitable damage to walls, curbs, etc., and the fact that repairs to walls and things of that kind cannot be carried out immediately due to the lack of labor."[47] Another common complaint, which Waterhouse neglected to mention, came from Britons who charged that their government was failing to provide just compensation for the property it seized.

NONFERROUS METALS

Steel was not the only metal of interest to Stettinius. Soon after he returned to the United States, he criticized the British for having made

[45] Ministry of Works and Planning, "Note on Augmenting the Supplies of Scrap Metal for War Production in Great Britain," n.d., in Stettinius, "Lend-Lease," Stettinius Papers, Box 149.

[46] Office of Lend-Lease Administration, British Meeting, 19 Aug. 1942, 3, Stettinius Papers, Box 137.

[47] George B. Waterhouse to Edward R. Stettinius Jr., 4 Sept. 1942, 3–4, NARA, Record Group 169, PI 29 9, Box 167.

"no real effort" to collect brass, bronze, and other nonferrous metals. Noting that Britain lacked the facilities to refine large quantities of nonferrous scrap, he expressed the hope that 40,000 to 50,000 tons might be collected and shipped to the United States for processing. "I stressed the desirability of this step to the British officials, and also to the Harriman Mission."[48]

The reason that the United States exerted so much pressure on the British to salvage nonferrous metals in autumn 1942 resulted from predictions that the Allies would face a deficit of 817,900 short tons of copper (20 percent of the 4.1 million short tons it required) in the coming year.[49] The copper situation was so serious that the United States took the unprecedented step of making the pennies it minted in 1943 entirely from steel.

On 17 August 1942, A. M. Baer, controller of nonferrous metals in the Ministry of Supply, chaired a meeting of officials from several other government ministries. Referring to Stettinius's allegations that they had done little to salvage nonferrous metals, he told his colleagues that because "the Americans were themselves short" of copper and zinc, "they needed convincing of the power of our Salvage Drive before being in the mood to supply this country." If voluntary contributions of items made from these metals did not prove sufficient, he warned that his majesty's government might use its powers of requisition. Before taking this "drastic" step, which would require "forcible entry etc." into people's houses, the committee decided to see whether a public relations campaign by the Ministry of Information would boost donations. "Mr. Baer said that the Appeal *must* go forward – there was no alternative as America was taking most strenuous action and we must conform."[50]

Meeting with British officials two days later, Stettinius "observed that at present no campaign was being carried on by the British for the collection of nonferrous scrap metal particularly copper and brass. Nonferrous Metal Control informed him that from 40 to 50 thousand tons of copper and brass could be collected if a campaign was started." In extremely blunt language, he predicted that "the fact that no campaign had been started for the collection of this nonferrous scrap might prove

[48] Stettinius, "Lend-Lease," 7.

[49] Combined Production and Resources Board, Interim Report, 14 Nov. 1942, cited in Copper: Interim Report No. 3, 27 Jan. 1943, NARA, Record Group 169, PI 29 9, Box 167.

[50] Notes of a meeting at the Directorate of Salvage and Recovery on 17 Aug. 1942, NA, AVIA 22/3088.

embarrassing to [the] British when they asked us to send them these materials."[51]

Stettinius's minutes of a subsequent meeting on 2 September 1942 note that he warned the British that unless their collections of nonferrous metals showed improvement, their claim on Lend-Lease funding could be jeopardized.[52] British sources corroborate this account. According to notes made by one of the Britons present, "Mr. Stettinius said he would like to emphasize again what appeared to him to be the necessity for a house-to-house campaign in Great Britain for the collection of scrap copper and scrap brass. He said that such a campaign was contemplated for the U.S. and that the repercussions in the U.S. would be most unfavorable if British householders were known to be retaining articles of a type which the U.S. householder had voluntarily surrendered to the war effort."[53] Five weeks later, Stettinius reiterated his displeasure with the state of recycling in the United Kingdom. As a member of the British Raw Materials Mission related in a cypher telegram to London, "Stettinius again referred specially to feeling in certain influential quarters in U.S. about inaction on part of U.K. on instituting serious campaign for collection of copper and brass scrap which he said had been estimated to amount to 50,000 tons."[54] Three days later, British officials informed Washington that they were setting up a nationwide campaign to encourage people to donate nonferrous metals for the war effort. In addition, they began compiling "lists indicating kinds of goods and stores and the order in which they should be taken over if requisition [is] ultimately decided upon."[55] To highlight the comprehensive nature of their approach, they informed the Americans that their plans included "door-handles, taps, name-plates, bells, candlesticks, ash-trays, coal boxes, fire screens, ornaments, small weights, and musical instruments like trumpets."[56]

51 Office of Lend-Lease Administration, Minutes of British meeting, 19 Aug. 1942, 3, Stettinius Papers, Box 137.

52 Office of Lend-Lease Administration, minutes of British meeting, 2 Sept. 1942, Stettinius Papers, Box 137.

53 "Stettinius Meeting of Wednesday, September 2, 1942," NA, T 160/1153.

54 British Raw Materials Mission in Washington to R.M.5. (A), 9 Oct. 1942, NA, AVIA 22/3088.

55 Cypher telegram for R. D. Fennelly, 12 Oct. 1942, AVIA 22/3088.

56 George B. Waterhouse to Edward R. Stettinius Jr., 9 Dec. 1942, NARA, Record Group 169, PI 29 9, Box 167.

In advance of the nonferrous scrap drive, the press office of the Directorate of Salvage and Recovery encouraged local and national newspapers to help publicize the campaign. On 6 October 1942, the directorate's press office issued a sample "editorial" that called upon the people of Britain to contribute nonferrous metals to the war effort:

> The need for these metals in our war factories and ship-building yards is urgent. In a battleship there are 2,000,000 lbs. of copper, brass is used for switches and other electrical apparatus, for compasses and navigating equipment, dials and gun-sights, and propellers are made of bronze, large quantities of zinc and brass go into cartridge cases. There are two miles of copper wire in a bomber, as well as brass for instruments, bronze for engine bearings, aluminium for fuselage and fittings, and anti-friction metal for engine bearings.[57]

Four days before the drive officially began, the Metropolitan Railway (part of the London Underground) announced plans to strip twenty trains of their ornamental bronze nameplates, including one that honored the nurse Florence Nightingale.[58] In addition to orchestrating other such publicity stunts, the Ministry of Supply made full use of the broadcasting facilities of the BBC and placed advertisements in eleven major newspapers.[59] Although the campaign was originally scheduled to last for just two weeks, the ministry extended its duration to a month and a half.

The British soon outpaced their North American allies in the range of objects they considered for the scrap heap. The same day that the drive began, G. B. Hutchings, principal director of salvage and recovery, met with half a dozen other government officials to discuss ways to find additional sources of nonferrous metals in Britain. In addition to searching government departments for every bit of nonferrous metal that could be spared, the officials declared that the government should consider melting down statuary. They agreed that "payment, whether as a purchase price or as compensation on requisitioning, should be closely related to the market value of the metal content of the statue. For this reason statuary with a definite market value as works of art should not be acquired in the first instance." They went on to discuss church bells, "which represent a very considerable quantity of N.F. [nonferrous] metal. Approach to be made to the leaders of the Churches, after consultation with the Ministry of Home Security. It is thought that bells of a date

[57] Ministry of Supply, Press Office, "Urgent Need for Unwanted Metal," 6 Oct. 1942, NA, AVIA 22/3088.

[58] "News in Brief," *Times* (London), 15 Oct. 1942, 2.

[59] For examples of these promotional materials, see NA, AVIA 22/3088.

earlier than 1700 A.D. should not be taken. In special cases, plaster casts of bells taken could be made, with a view to ultimate replacement. It is appreciated that such a step might arouse considerable opposition in the country on religious and psychological grounds." Bells deemed "irreplaceable . . . ought not to be requisitioned, except in dire national emergency."[60] In the end, the British government never took the step that many countries did during the war of melting down large numbers of church bells and public statuary. Doing so might well have caused public outrage; just a year earlier, a British magazine story about German plans to scrap church bells in occupied Norway had carried the headline "Nazi Vandals."[61]

Such rhetoric was a frequent part of wartime news coverage, which frequently depicted British salvage efforts, no matter how extreme, as mild compared to those of their German enemies. Referring to "the vast totalitarian drive now in progress on the German home front," the *Times* reported in summer 1942 that Albert Speer, the minister for armaments and munitions, had issued a decree that subordinated absolutely everything to the demands of weapons production. "Even modern machinery and whole industrial plants in so far as they are not adaptable for immediate use in armaments production are to be scrapped without compunction," observed the article's author. "Valuable industrial plant in perfect condition, idle only because of the war, will be destroyed and its owners dispossessed. That this constitutes the destruction of immense capital values and heavy loss of industrial potential is not denied. On the contrary, it is declared officially that 'all usual financial considerations and regard for post-war production are to be left out of account.'"[62] In October 1942, four months after they renamed their journal to reflect the growing importance of recycling, the editors of what was now *Public Cleansing and Salvage* declared, "In Germany scarcely a door handle has been spared in the round-up of metal for the prosecution of the war. . . . Hitler spares nothing and no one. Here, though the need for metal is equally great (we are consuming more than 1,000 tons an hour), requisitioning has been much less drastic." Regardless of whether Britain adopted more draconian means, the editorial asserted, "Our scrap metal

[60] G. B. Hutchings, Notes of a meeting on 19 Oct. 1942, 2 Nov. 1942, NA, AVIA 22/3088.

[61] "Nazi Vandals," *Waste Trade World and the Iron and Steel Scrap Review* (hereafter *WTW*), 23 Aug. 1941, 3.

[62] "The Speer Plan in Action: Mobilizing Reserves of Men and Materials," *Times* (London), 7 Sept. 1942, 5.

collection must be 100 per cent. efficient, and the more determinedly we seek to make it so, the sooner will the war be won."[63]

In October 1942 Stettinius asked Harriman to provide frequent reports on the progress of Britain's nonferrous metals drive, "as I feel it is a very delicate question and one of great importance. We should have current facts on it here in case we ever have to answer questions on the matter."[64] Two weeks later, Stettinius told a British official that he wanted to be in a position to tell the House and Senate appropriations committees "that the United Kingdom is collecting non-ferrous scrap along the same lines that have been adopted by Canada and the United States."[65] In their internal discussions, British officials frequently recognized the importance of doing everything possible to persuade reluctant U.S. politicians to support Lend-Lease. In January 1943, for example, an official in the Board of Trade expressed concern when the "extremely well informed Washington correspondent" of the London *Times* reported significant opposition in Congress to further Lend-Lease appropriations. "I do not doubt," asserted the British official, "that Mr. Stettinius will be hoping for an answer . . . which will strengthen his hand in replying to Congressional criticism. In these circumstances it is, I think, of the first importance that we should go as far as possible to meet him."[66]

Bowing to heavy U.S. pressure, in August 1943 the British government announced that anyone holding more than one ton of copper, lead, or zinc had to report it.[67] Britain's top official in charge of nonferrous metals, A. M. Baer, told his colleagues in the Ministry of Supply that they needed to "hurry somewhat over this Order or we shall have the Americans on our tails again."[68] In another move designed to assuage Washington, the Directorate of Salvage and Recovery provided U.S. officials with a "comprehensive selection of posters and literature" that the British were using to promote recycling. These achieved their intended effect. According to M. A. Colebrook, vice counsel at the American embassy in London, this material demonstrated that the British were

[63] "Editorial," *Public Cleansing and Salvage*, Oct. 1942, 44.

[64] Edward R. Stettinius Jr. to Averell Harriman, 19 Oct. 1942 (copy), NARA, Record Group 169, PI 29 9, Box 167.

[65] Edward R. Stettinius Jr. to Clive Baillieu, 2 Nov. 1942 (draft copy), NARA, Record Group 169, PI 29 9, Box 167.

[66] Board of Trade, Minute Sheet, 11 Jan. 1943, NA, BT 11/2078.

[67] A. M. Baer to George W. Meek, 5 Aug. 1943 (copy), NA, AVIA 22/3088.

[68] A. M. Baer to T. Turner, 5 Aug. 1943 (copy), NA, AVIA 22/3088.

"intensively and intelligently" working to salvage all they could for the war effort.[69]

EMPIRE

American officials paid a great deal of attention not only to Great Britain but also to its vast empire. In keeping with their aversion to imperialism (at least when practiced by others), many Americans were wary of Churchill's frequent assertions that his country was fighting to save the British empire. Lend-Lease officials, anxious to demonstrate that American aid promoted freedom and democracy rather than imperialism, sometimes used their influence to counter what they viewed as British misrule of its empire. On 19 December 1941, Winthrop G. Brown, deputy director of the Harriman Mission in London – a man who in the 1960s would serve as U.S. ambassador to Laos and later South Korea – informed Stettinius that

> people in the Dominions and Colonies see the British Isles being supplied under Lend-Lease with many things which are denied to them, and they cannot help feeling that since all requests go through the BPC [British Purchasing Commission] this is in large part a British decision and that they are not being given fair and proper consideration. They feel they are just as much at war as the people of the British Isles and if they got more nearly similar treatment under Lend-Lease, a lot of oil would have been put in the machinery of the Empire war effort.[70]

Three weeks later Brown wrote again to provide his assessment of conservation efforts in the British Empire. His observations reveal a great deal about the kind of behavior that Washington expected from recipients of U.S. aid. "I see daily evidence," he told Stettinius, "that the Dominions are tightening their belts and exercising more and more strict controls over their use of critical materials, particularly steel."[71]

Although most British officials assumed that decisions about what to do with the natural resources of Britain's colonies should be made in London, a far different attitude prevailed in many other parts of the

69 M. A. Colebrook, "The Salvage of Household Scrap in Great Britain – 1943," 28 Dec. 1943, NARA, Record Group 169, A 1 500, Box 803, File 77821.

70 Winthrop G. Brown to Edward R. Stettinius Jr., 19 Dec. 1941, NARA, Record Group 169, PI 29 9, Box 164. For consistency, I have taken the liberty of adding hyphens to Lend-Lease in this quotation, as that is how his other letters spell it. The anomaly likely resulted from the work of a typist who was new to the job.

71 Winthrop G. Brown to Edward R. Stettinius Jr., 10 Jan. 1942, NARA, Record Group 169, PI 29 9, Box 164.

world, including Washington. Writing from the U.S. capital in April 1942, a British official informed his colleagues in London that their actions struck many in the colonies as overbearing and were making "it difficult for us to secure wholehearted cooperation of Empire Ministers and heads of Supply Missions." As bureaucratic telegrams go, this one was the equivalent of a shouted warning. "There are obvious dangers," it continued, "in our attempting to bring influence to bear on Home Govts. through Missions here in view of what we understand to be general tenor of instructions so far received by Empire Missions." The latter, explained the telegram, were "unlikely to be willing advocates of policy of routing everything for CRMB [the Combined Raw Materials Board] through London [and] if pressed there is definitely risk that they will advise their Home Govts. in a sense contrary to your wishes." This telegram raised the specter that, as had happened during the First World War, participation in a global conflict would encourage and accentuate feelings of independence among colonized people. In addition, the telegram suggested the prospect that Britain's colonial satellites would begin to orbit around Washington instead of London.

In light of the ascendant economic and strategic power of the United States, some British officials suggested that London needed to do more to persuade its colonies and dominions that they benefited from their connection with Britain. If they failed, it seemed probable that the United States might supplant Britain's imperial position – and the telegram suggested that this was already starting to occur. Its author explained that officials in the Office of Lend-Lease Administration "require and are obtaining from Empire Missions information about overall picture of supplies and requirements for each material . . . therefore building up by direct contact with home govts. [a] body of information for many materials." The telegram ended with a prescient warning: an effort by London to reassert its authority "will not be understood by Missions here and may well lead to renewed pressure for direct representation on combined Board with consequences difficult to foresee."[72] The United States, in other words, threatened to erode the bonds that tied Britain's colonies to their "mother country." At a meeting with American officials in August 1942, Sir John Duncanson of the Iron and Steel Control noted that the United States now supplied most of the imported goods that India

[72] Cypher telegram from British Raw Materials Mission in Washington to London, 8 Apr. 1942, NA, T 160/1153.

and the rest of the British empire required.[73] Sensitive to U.S. hostility toward imperialism, a senior British official argued that it was essential to "convince sections of the U.S. public that our colonies are not run like the private estate of a landlord. We should explain the principles on which our colonial policy has been founded and how we have consistently applied liberal ideas for the benefit of the native peoples."[74]

After the U.S. announcement in August 1941 that it would soon stop exporting steel scrap, the British intensified their search for scrap metal throughout their empire. Many ships and crews were lost attempting to carry this vital cargo. One such vessel was the SS *Nagpore*, which the British government contracted to transport scrap metal from Kenya to the United Kingdom in 1942. The East Africa Provision Office, which assembled its cargo, had until this point made a practice of sending ferrous scrap to Britain and nonferrous scrap to India. On this occasion, however, dockworkers loaded the ship with both types of metal, which had been gleaned from across Somalia and Kenya. On 13 May the *Nagpore* left Mombasa laden with 98 tons of spring steel and 1,215 tons of miscellaneous scrap, which included aluminum, brass, nickel steel, lead, cast iron, 38 tons of nuts and bolts, and 5 tons of machine gun parts. In peacetime, ships sailing between the Indian Ocean and the United Kingdom often shortened their journeys by traveling via the Suez Canal and the Mediterranean. To avoid the wartime hazards of the Mediterranean, however, the ship's captain settled on the much longer route clockwise around the African continent. On 31 October, just four days before the *Nagpore* was expected to arrive in Manchester, a port transportation officer informed the Iron and Steel Control that the *Nagpore* "is a casualty, no details are available, but I understand she is not now likely to report."[75]

Africa was an important source not only of scrap but also of mineral ores and rubber. In April 1942, Harold Macmillan, who had recently moved to the Colonial Office, wrote to Sir Andrew Duncan, the minister of supply, about the declining rate of tin production in Nigeria. Macmillan noted that "we have been rather concerned here at the attitude of some of the tin producers towards present urgent needs. We have the

[73] Ministry of Production, Empire Clearing House, Minutes, 26 Aug. 1942, NA, T 160/1153.

[74] "Memorandum on Colonial Policy," 18 Jan. 1943, Cherwell Papers, Nuffield College Library, Oxford, F.401/5. Extracts from the Cherwell Papers are reproduced with the kind permission of the Cabinet Office Historical and Records Section.

[75] Director of Salvage (War Office) to Deputy Director of Economy (Disposals), 4 June 1942, NA, AVIA 22/692; Assistant Port Transportation Officer (Manchester) to Iron and Steel Control, 31 Oct. 1942, NA, AVIA 22/692.

FIGURE 14. The global reach of Britain's wartime salvage efforts. During the war British officials organized large-scale salvage drives throughout the British Empire. This photo, taken in Bechuanaland (now Botswana) in 1942, shows villagers contributing their iron kettles to the Allied war effort.
Source: Ministry of Information Second World War Official Collection. Courtesy of Imperial War Museum, © IWM (K 2496).

feeling, which I believe is shared in your Department, that some of these Companies are not pulling their weight." Contradicting himself, however, Macmillan went on to acknowledge that British efforts to extract tin from its colonies were so intensive that they had reached "uneconomic and destructive" levels.[76] In April 1943, a British cabinet minister credited Africans with contributing significant quantities of copper ore, bauxite, and iron ore to the Allied war effort. He further noted that shipments of rubber from Africa had increased fourfold over the previous year.[77]

Although the Allies had controlled almost all of the world supply of rubber at the start of the war, this situation changed when Japan invaded Malaya and the Dutch East Indies.[78] On 14 January 1942, three days

[76] Harold Macmillan to Andrew Duncan, 14 Apr. 1942, NA, AVIA 11/10.
[77] "British Ties with U.S.," *Times* (London), 9 Apr. 1943, 8.
[78] Ministry of Supply, "Memorandum on British Methods of Raw Material Conservation," Sept. 1943, 11, NA, AVIA 46/490; Alfred Plummer, *Raw Materials or War Materials?* (London: Victor Gollancz, 1937), 13, 16.

after Japanese troops entered Kuala Lumpur, the *Times* observed that the Japanese occupation of Malaya made it essential to reclaim rubber. Almost anything made of rubber could be recycled: old tires, water bottles, sponges, boots, garden hoses, and aprons.[79] The good news was that Britain had a stockpile of 100,000 tons of old rubber that people had donated since the war began. The bad news was that the country could recycle only 12,000 tons of rubber a year.[80] A short time later, in a memo stamped secret and confidential, Stettinius informed Harriman that the recent developments in Asia threatened the United States as well. "The general rubber supply situation looks extremely serious," he explained. Even if all civilian consumption were eliminated, "the stockpile might be exhausted by the end of 1943."[81] In response to a request from Harriman, the British soon exported 10,000 tons of scrap rubber for recycling in the United States.[82]

In March 1942, the Foreign Office telegraphed the British Raw Materials Mission in Washington with information, intended for public dissemination in the United States, about efforts to consume rubber in the United Kingdom. In addition to noting that the use of rubber in nonessential products was now illegal in Britain, it asserted that the purchase of both new and used tires was strictly controlled. "Rationing," it argued, was "considerably more drastic than [the] present American scheme." Efforts had begun to recycle rubber from old tires and other articles, but in the terse words of the telegram, "Reclamation frankly still in infancy."[83]

Even though the British were unable to recycle all their old rubber, they nonetheless found ways to reuse it. One of the most important of these was in the construction of runways. In 1941 alone the Air Ministry used an estimated 40,000 tons of tire chippings for this purpose.[84] Another use for old tires was in diversionary fires designed to lure German air crews

79 "'Turn Out Old Rubber': National Campaign," *Times* (London), 16 Jan. 1942, 2.

80 W. L. Clayton to Lord Beaverbrook, 11 Jan. 1942 (copy), NARA, Record Group 169, PI 29 82, Box 681.

81 Edward R. Stettinius Jr. to Averell Harriman, 18 Feb. 1942, NARA, Record Group 169, PI 29 83, Box 685.

82 "Notes on Mr. McCabe's 9:00 a.m. Staff Meeting," 10 Feb. 1942, Box 137, Stettinius Papers.

83 Cypher telegram from Foreign Office to British Embassy in Washington, 29 Mar. 1942, NA, T 160/1153.

84 Ministry of Supply, Salvage Department, "Conference of Honorary District Advisers Held at the Town Hall, Leicester, on Saturday, 20th. September, 1941," Shakespeare Centre Library and Archive, Stratford-upon-Avon, BRR 55/14/31/4/1.

into dropping their bombs on unpopulated areas outside of cities.[85] In April 1942, Hutchings contacted the Women's Department of the Trades Union Congress (TUC) and the National Federation of Women's Institutes to ask for their help in stimulating the search for scrap rubber. Alluding to the Japanese invasion of Malaya, he noted that "in view of developments in the Far East this matter is of paramount importance."[86] A short time later the government launched a series of county-level campaigns to collect rubber scrap.[87]

The question of rubber conservation and salvage was intertwined with automobile use and the rationing of fuel. In July 1942, an official in the British embassy in Washington reported that White House economic adviser Dr. Isador Lubin had told him that the effort of the U.S. government to encourage its citizens to recycle rubber was "a political move necessary to discredit certain interests who have stated that scrap is available in sufficient quantities to make petrol rationing unnecessary as a rubber conservation move. It is expected that once the scrap drive has been finally shown to have been a failure, the President will make a frontal attack on the rubber problem through nation-wide petrol rationing."[88]

In early August 1942, the Ministry of Supply informed the Harriman Mission that the people of Britain had contributed over 68,000 tons of scrap rubber to the war effort during a six-month period ending in July 1942.[89] Rubber collections in Britain during 1942 totaled 115,000 tons, considerably more than the 75,000 tons the Ministry of Supply had hoped to collect.[90] Despite these impressive achievements, British officials feared that supplies of rubber for essential needs would soon run low, especially if the anticipated manufacture of large quantities of synthetic

85 Harold Macmillan to William Palmer, 15 Oct. 1941 (copy), Harold Macmillan Papers, Department of Special Collections and Western Manuscripts, Bodleian Libraries, University of Oxford, c. 273, fol. 177.

86 G. B. Hutchings to Frances Farrer, Apr. 1942 [received 14 Apr.], Records of the National Federation of Women's Institutes, Special Collections and Archives, London School of Economics, 5FWI/A/3/073; G. B. Hutchings to Miss Adams, Apr. 1942 [received 14 Apr.], Trades Union Congress Collection, Modern Records Centre, Warwick University, MSS. 292/557.61/1.

87 Directorate of Salvage and Recovery, Campaign for the Recovery of Rubber Scrap, Apr. 1942, Trades Union Congress Collection, Modern Records Centre, Warwick University, MSS. 292/557.61/1.

88 British Embassy in Washington to Foreign Office, 15 July 1942 (copy), NA, BT 28/961.

89 Arthur Smith to Arthur Notman, 5 Aug. 1942 (copy), Stettinius Papers, Box 149.

90 Ministry of Supply, "Memorandum on British Methods of Raw Material Conservation," Sept. 1943, 12, NA, AVIA 46/490.

rubber in the United States ran into difficulties.[91] On 7 September 1942, two years to the day after the start of the Blitz, the Ministry of Supply made it illegal, under the Salvage of Waste Materials (No. 4) Order, to destroy or discard waste rubber.[92]

[91] "Memorandum for Lord Cherwell on the Rubber Position," 16 Nov. 1942, Cherwell Papers, Nuffield College Library, Oxford, H.59/1–5.

[92] The Salvage of Waste Materials (No. 4) Order was issued as S.R. & O. 1942 No. 1770. See Great Britain, *Statutory Rules and Orders 1942*, vol. 2 (London: HMSO, 1943), 1993. For a discussion of this order, see "Salving of Rubber," *Times* (London), 7 Sept. 1942, 2.

PART III

HISTORY, CULTURE, AND CIVIL LIBERTIES

7

The Built Environment

> Come on! set fire to the library shelves! Turn aside the canals to flood the museums! . . . Oh, the joy of seeing the glorious old canvases bobbing adrift on those waters, discolored and shredded! . . . Take up your pickaxes, your axes and hammers, and wreck, wreck the venerable cities, pitilessly!
>
> – F. T. Marinetti, 1909[1]

On 2 August 1940, Winston Churchill issued a typically terse note asking his chief of staff, General Hastings "Pug" Ismay, to tell him "what has been done about the collection of scrap of all kinds. Let me have a short report on one page covering the progress made this year."[2] Eight days later, the Ministry of Supply sent Churchill a summary of the country's scrap collection efforts. It noted that the value of materials collected and sold by local councils had jumped from £37,000 in December 1939 to £205,000 in June 1940. Monthly sales of reclaimed paper had increased fivefold, to 24,000 tons; metals fourfold, to 28,000 tons; and kitchen waste tenfold, to 3,000 tons. Of the 984 communities that were required to establish recycling collection programs, all had submitted their plans for doing so. In addition to diverting recyclable materials from the municipal waste stream, the ministry had initiated "extensive campaigns for

[1] F. T. Marinetti, "The Founding and Manifesto of Futurism" (1909), in R. W. Flint, ed., *Marinetti: Selected Writings*, trans. R. W. Flint and Arthur A. Coppotelli (London: Secker & Warburg, 1972), 43.

[2] Winston Churchill to Hastings Ismay, 2 Aug. 1940 (extract), NA, CAB 120/388 (also in PREM 3/378).

scrap iron from factories and works, estates and farms."[3] Ministry of Supply officials frequently asserted that to make up for the decline in imported raw materials, the nation needed to look beyond traditional sources. The manner in which these campaigns unfolded added considerably to the wartime destruction of Britain's material culture.

HISTORICAL ARTIFACTS

Soon after the war broke out, officials in the industrial town of Middlesbrough announced that they were donating as scrap several pieces of artillery that British forces had captured from Germany during the Great War.[4] A few months later a school in Edinburgh contributed a German howitzer of similar vintage to the scrap metal drive.[5] J. L. Henderson, an Iron and Steel Control official within the Ministry of Supply, suggested in May 1940 that "old guns, tanks, trophies and other relics, which had only sentimental value, should be donated by their owners."[6] During the four months that followed, people across Britain scrapped 800 historic field pieces, including several that had long been displayed at the Tower of London.[7] One of the first to respond to the call was Countess Roberts, who contributed cannon and field guns from the campaigns of her late father, Field Marshal Frederick Roberts, who had fought for Britain from the Indian Rebellion of 1857 to the Second Boer War.[8] Authorities in Salford gave up two cannons from the Siege of Sevastopol that had stood in the city's Peel Park since the Crimean War, along with their accompanying cannonballs.[9] In addition to old artillery pieces, people scrapped a wide range of other objects. Manchester residents contributed "a battered trumpet, blown by Queen Victoria's State trumpeter nearly a century ago," and a German cavalry sword that had been dropped "during the Hun retreat

[3] Ministry of Supply, "Memorandum on the Progress of Salvage Work during 1940," n.d. [Aug. 1940], NA, CAB 120/388 (also in PREM 3/378).

[4] "Government Salvage," *Waste Trade World and the Iron and Steel Scrap Review* (hereafter *WTW*), 25 Nov. 1939, 3.

[5] "Edinburgh's Campaign," *Public Cleansing*, Apr. 1940, 259.

[6] "The Institute of Public Cleansing: Scottish Centre at Glasgow," *Public Cleansing*, June 1940, 314.

[7] "Old Guns for the Melting Pot," *WTW*, 10 Aug. 1940, 8; "Collecting Old Iron and Steel," *Times* (London), 15 Aug. 1940, 9; "Tower Guns for Scrap," *WTW*, 24 Aug. 1940, 2.

[8] "Arms Drive for Victory," *Times* (London), 1 June 1940, 3.

[9] "In Brief," *Manchester Guardian*, 7 Aug. 1940, 6.

of 1918."[10] Another object to be sacrificed was "a fine old church bell, dated 1680 and inscribed 'James Bartlett made me.'" Despite its antiquity and good condition, the bell was soon melted down to aid the war effort.[11]

Amid all this scrapping, those who sought to preserve historic artifacts achieved some victories. In August 1940, for example, officials in the Ministry of Supply declined an offer from the town council of Falmouth to sacrifice the guns from the wooden frigate *Bellerophon*, aboard which Napoleon had surrendered after Waterloo, and "the 80 cwt. [hundred-weight] anchor" of an old naval training ship, the *St. Vincent*. This anchor, noted a contemporary report, was "a unique example of water-hammer forging, and is the only one of its kind in the country."[12] At roughly the same time, an official at the Tower of London revealed that "some of the old guns recently dug up in connection with the campaign for scrap metal are older than first thought. Instead of coming from the period of the French and Napoleonic Wars, they predate 1714. Twelve of these guns will be preserved." According to one reporter, the reason that so few guns from this period remained was because most had been melted down in the early nineteenth century in "a nation-wide scrap campaign organised by [Prime Minister William] Pitt to help in the defeat of Napoleon."[13]

On 24 September 1940, the Ministry of Supply launched a "National Survey of Fixed and Demolition Scrap." The purpose of the survey was to compile an inventory of all "old buildings, mines, bridges, machinery, railway track, and a host of derelict structures containing thousands upon thousands of tons of much-needed scrap metal." Experts predicted that the survey would add half a million tons of steel to the nation's arsenals – enough metal to build 7,000 heavy tanks.[14] Although the authorities promised to preserve objects of historical importance, opinions differed as to what ought to be saved. Artifacts from the nineteenth century struck many as too recent to qualify. In the spring 1941, for example, officials

[10] "Scrap for Victory," *WTW*, 13 July 1940, 2.

[11] "Sidelights on the Scrap Front," *WTW*, 31 Aug. 1940, 22.

[12] "Naval Relics Saved," *WTW*, 24 Aug. 1940, 2. An imperial hundredweight consists of eight stone, which total 112 lbs.

[13] "Marlborough Guns," *WTW*, 7 Sept. 1940, 2.

[14] "National Survey of Fixed and Demolition Scrap," *Builder*, 11 Oct. 1940, 367; "The National Survey of Fixed and Demolition Scrap Iron and Steel: Results of First Month's Test Operation," *Public Cleansing*, Dec. 1940, 118; G. D. N. Worswick, "British Raw Material Controls," *Oxford Economic Papers* 6 (Apr. 1942): 1–41, quoting 5.

TABLE 4. *Inputs of Scrap to the UK Steel Industry, 1941–43 (long tons)*

	Sept 1939–30 Mar 1941	31 Mar 1941–31 Dec 1941	Jan 1942–June 1942	July 1942–Jan 1943
Imports	1,285,921	402,878	1,812	none
Industry and scrap dealers	11,821,756	7,097,824	5,194,432	4,777,068
Buildings damaged by air attack	68,702	340,675	139,432	66,804
Railings	34,900	18,620	222,363	187,466
Household and municipal scrap	76,600	180,632	149,217	161,400
Tram rails	33,000	20,500	16,219	16,560
Village dumps[a]	38,186	2,214	n/a	n/a
Obsolete buildings/ industrial equipment[a]	824,298	188,592	n/a	n/a
Obsolete farm equipment[a]	1,500	1,033	n/a	n/a
National Survey scrap	n/a	n/a	219,689	358,556

[a] From January 1942, these categories were combined into a new category called "National Survey" Scrap.

Sources: M. A. Colebrook, "British Iron and Steel Scrap Collections: September, 1939–December 31, 1941," 2 March 1942, NARA, RG 169, A 1 500, Box 1670, File 412841; M. A. Colebrook, "British Iron and Steel Scrap Collections: January 1, 1942–June 30, 1942," 27 July 1942, NARA, RG 169, A 1 500, Box 1907, File 484851; George B. Waterhouse to Edward R. Stettinius Jr., 4 Sept. 1942, NARA, RG 169, PI 29 9, Box 167; M. A. Colebrook, "British Iron and Steel Scrap Collections: July 1, 1942–January 2, 1943," 6 Feb. 1943, NARA, RG 169, A 1 500, Box 1838, File 459912.

scrapped two Boulton and Watt steam engines from the Deptford Pumping Station, one installed in 1812 and the other in 1824.[15]

The search for heavy iron scrap also destroyed the last remaining pieces of the Crystal Palace. Designed by Joseph Paxton to house the Great Exhibition of 1851, the Crystal Palace owed its name to its glass walls and roof. After the exhibition closed, workers painstakingly dismantled the palace from its original location in Hyde Park and re-erected it near Sydenham, south of London. There it stood for the better part of a century. In November 1936, materials being stored in the building caught fire, and the intense heat softened the ironwork to the point that most of

[15] "Vintage Pumping Engines for Scrap," *WTW*, 7 June 1941, 5.

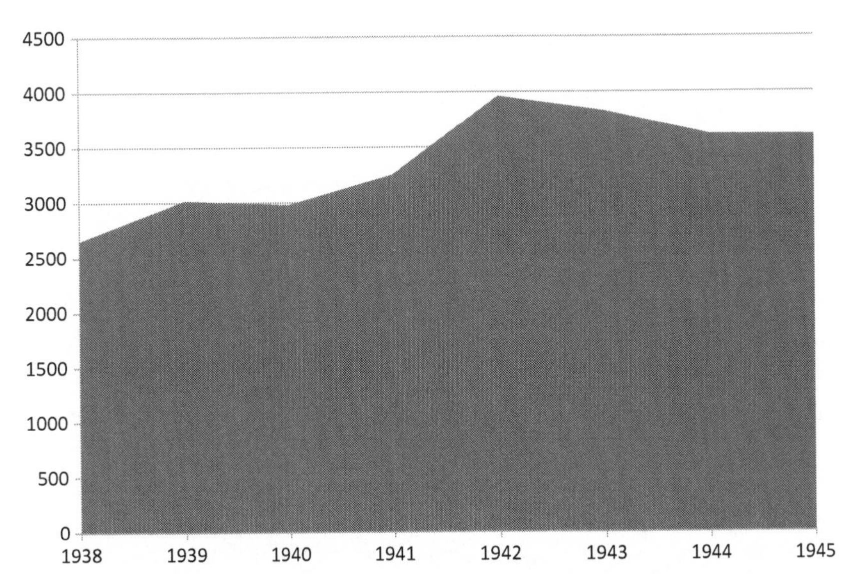

FIGURE 15. Iron and steel scrap purchased within the United Kingdom (thousand tons), 1938–45.
Source: Peter Howlett, *Fighting with Figures: A Statistical Digest of the Second World War* (London: Central Statistical Office, 1995), 6.5.

the structure collapsed. All that remained standing were its two enormous water towers, designed by the renowned engineer Isambard Kingdom Brunel to supply the building's enormous decorative fountains. Each tower was approximately 280 feet tall and contained 800 tons of iron. Without any public discussion about whether Brunel's creations ought to be preserved, the government contracted a scrap firm to demolish both towers in the early months of 1941.[16] Even though the last remnants of the structure disappeared long ago, people still refer to the area where it stood as Crystal Palace.

Shortly before Brunel's towers came down, a debate took place in the West Midlands town of Walsall over a town landmark known as the Bridge Clock, which some thought should be scrapped in support of the war effort. According to the proposal's sponsor, "The clock would have to go sooner or later and there could not be a better time for its removal than the present, as the metal contained in the standard could be applied

[16] "End of Palace Tower," *WTW*, 29 Mar. 1941, 2–3; "Crystal Palace North Tower Felled," *WTW*, 19 Apr. 1941, 9; "End of Crystal Palace North Tower," *WTW*, 26 Apr. 1941, 4–5.

to a most important purpose. It was estimated that the clock and its support contained from three to four tons of 'scrap,' and if the Council did not surrender it voluntarily, the Ministry of Supply might come along later and schedule it for compulsory demolition in the national interest." Others argued that the clock not only continued to serve a utilitarian function but that it should be saved for "sentimental" reasons. After nearly an hour of debate, more than half of the members of the town council voted to save it.[17]

One of the most articulate campaigners for historic preservation in wartime Britain was the indefatigable historian and archivist Joan Wake, who had a reputation for traveling to remote country archives not by rail or automobile, but on the seat of a motorcycle. She carried her best hat with her in a box strapped to the machine. "Last week," she wrote to the *Times* in August 1941, "I heard a broadcast talk in which the narrator described with great pride and satisfaction the removal as scrap iron from a west country town of four old guns of 1789." Pointing out that "there is no one centrally responsible for the care of our national treasures and historical records," Wake asked, "Is it not time that someone in authority curbed the enthusiasm of the salvage department? Is it really in the best interests of the nation that such things should be sacrificed before other sources of supply have been exhausted?"[18] Responding to her letter, the master of armories at the Tower of London admitted that "the destruction of old guns for scrap-iron . . . has undoubtedly been extensive during the last 18 months," but he defended the War Office and the Ministry of Supply against the charge that they had acted wrecklessly. "The danger," asserted James G. Mann, "has come from misplaced zeal among local authorities and individuals, who do not pause to obtain expert advice before calling on the salvage contractor."[19]

The destruction of historical artifacts intensified as the war continued. At a meeting of regional salvage officials in the West Midlands in October 1941, one of them complained that local authorities sometimes collected "metal articles (antiques etc.)" and sold them to a scrap merchant, only to discover later that they "had been resold in their original form instead of being disposed of as scrap metal." Another representative at the meeting, John C. Jorden of the Birmingham Salvage Department, suggested that

[17] "The Bridge Clock: Proposal to Sell It for War Scrap," *Walsall Observer*, 10 Jan. 1941.

[18] Joan Wake, letter to the editor, *Times* (London), 22 Aug. 1941, 5; Peter Gordon, "Wake, Joan (1884–1974)," in *Oxford Dictionary of National Biography* (Oxford: Oxford University Press, 2004).

[19] James G. Mann, letter to the editor, *Times* (London), 26 Aug. 1941, 5.

"the best solution was for the articles to be broken up with a hammer immediately," which would destroy their value as artifacts and reduce them to their constituent materials.[20] Shredding, of course, had the same effect on books and manuscripts.

Among the countless items that British citizens donated for salvage during the war was a bronze plaque that commemorated a soldier killed in the First World War. The donor was the man's sister. Seemingly unaware of the irony of war memorials being turned into weapons, she reportedly hoped "that others will follow her example. Much valuable salvage would thus reappear in the form of guns and munitions."[21] That same year, an eighty-five-year-old man donated "a bag of some 400 old and mostly foreign coins weighing about 8lb., as a gift to the salvage drive."[22] Presumably these were to be melted and made into munitions, although if sold to collectors they might have contributed a great deal more value to the war effort.

IRON RAILINGS

Many people in wartime Britain viewed railings as unnecessary, ugly, and elitist. As one official with the Iron and Steel Control put it, "Few of the railings set in place during the past century have any æsthetic or artistic value." Reflecting a widespread disdain for Victorian style, he asserted that railings were nothing but "florid adornment" and "an empty sign of affluence."[23] Such hostility to railings was nothing new. In 1926, a contributor to the *Sunday Times* charged that Londoners were "strangely addicted" to railings, which caused "arbitrary and unnecessary material restrictions" to the enjoyment of their city.[24] A short time later the *Graphic* published an article suggesting that London would become more beautiful if it dispensed with its railings.[25] Soon after the Second World War broke out, those who sought to rid Britain of railings realized that the emergency

[20] Ministry of Supply, Salvage Department, District No. 3, minutes from a meeting in Birmingham, 1 Oct. 1941, 5, Shakespeare Centre Library and Archive, Stratford-upon-Avon, BRR 55/14/31/4/1.

[21] "Memorial Plaque for Salvage," *Public Cleansing*, Apr. 1942, 224.

[22] "40,000 Tons of Scrap a Week: National Drive Launched," *Times* (London), 22 Jan. 1942, 2.

[23] David Murray, "Railings," *Public Cleansing*, Jan. 1941, 162–3.

[24] "Railings: A National Penchant," *Sunday Times* (London), 6 June 1926, 13.

[25] Harold Cox, "London's Beauty behind Iron Bars: A Plea for the Removal of Our Useless Railings," *Graphic*, 21 Aug. 1926.

might help them achieve their aims. Within days of Hitler's invasion of Poland, a chorus of urban residents urged the removal of iron railings from London and other British cities. In a letter to the *Times*, one of them complained that railings around parks and squares constituted a particular hazard in wartime because they impeded access to the rudimentary air raid shelters that the government had excavated.[26] Others suggested that railings ought to be scrapped because they symbolized upper-class wealth and privilege that had little place in an egalitarian Britain that was fighting a "people's war." The removal of railings, argued Wilfred Leon in a letter to the *Times* a week after Britain declared war, would "be an act of democracy long overdue."[27]

London resident Peter Lewin agreed, asserting, "The cast iron horrors which defile our squares and public places must have few defenders. . . . Apart from the vandalism of enclosing a beautiful garden with a hideous, soot-laden palisade, there is the question of denying the casual wayfarer the opportunity of enjoying well-tended flower beds and the shade of pleasant trees." The best thing to do with railings, he argued, would be to fire them "forthwith into the Greater Reich."[28] Another observer declared that despite a long life spent in London, he had never seen a single person in Berkeley Square, which like many private squares was kept under lock and key. Echoing arguments that urban reformers had made since the nineteenth century, he expressed indignation that even in the heat of summer, "when the well-to-do in the great squares" fled London for their holidays, the less fortunate children of their neighborhood were forced to play in the streets because the squares remained off limits.[29]

In the spring of 1940 an official from the Ministry of Supply informed Walsall's town clerk, "Many iron railings, bollards and refuge posts in and around parks and gardens and in our streets have no important aesthetic value, and are not serving so essential a purpose that they may not be removed for scrap." The ministry went on to ask town councils to hand over all railings that they owned, with exception of any that were

[26] D. D. B., letter to the editor, *Times* (London), 11 Sept. 1939, 4.
[27] Wilfred Leon, letter to the editor, *Times* (London), 20 Nov. 1939, 4.
[28] Peter Lewin, letter to the editor, *Times* (London), 27 Apr. 1940, 7.
[29] Lionel Earle, letter to the editor, *Times* (London), 1 May 1940, 9. On earlier debates about access to London's squares, see Peter Thorsheim, "Green Space and Class in Imperial London," in *The Nature of Cities: Culture, Landscape, and Urban Space*, ed. Andrew C. Isenberg (Rochester: University of Rochester Press, 2006), 24–37.

"clearly needed for public safety."[30] In an attempt to convince members of the public to give up their railings, the ministry enlisted the well-known architect Sir Giles Gilbert Scott, designer of the iconic red telephone box as well as the Battersea and Bankside power stations (the latter is now the Tate Modern). Speaking at the opening of a month-long exhibition in London called Railings for Scrap, Scott asserted that very few of the railings created during the past hundred years warranted preservation.[31]

The reason that the Ministry of Supply targeted the railings of parks was not only because of their substantial metal content but also because of their prominence. In his notes of an April 1940 conversation with a member of the Iron and Steel Control, another civil servant recalled that the man was "delighted at the prospect of having park railings, + he is not so much concerned as to whether it be 5 tons or 50 tons as with the fact that the Controller can use this gesture in publicity work." This could cut both ways, however. Sir Patrick Duff, a top official in the Office of Works, advised, "You should warn this gentleman that he must be careful not to overreach himself over propaganda. The First Commissioner [of Works] considers that it would have a most heartening effect *in Germany* to hear that our Royal Parks are being denuded of all their necessary protection in order to eke out our supplies of scrap."[32]

Ruth Dalton, who chaired the London County Council's Parks Committee, welcomed the wartime push for novel sources of scrap, for she had "always felt that the beauty of London's parks could be greatly enhanced by getting rid of unnecessary iron railings."[33] By the end of June 1940 she had overseen the removal of 345 tons of railings from thirty-two parks and open spaces, including 40 tons from Battersea Park, 68 tons from Hackney Downs, and 44 tons from Hilly Fields in Lewisham.[34] In the

[30] Iron and Steel Control to town clerk, 24 July 1940, Walsall Local History Centre, 235/4. A slightly different form of this letter, which replaced the reference to munitions production with a broad statement about the need to "meet any contingency that may arise," went to other councils. See Iron and Steel Control to Middlesex County Council, 31 July 1940, London Metropolitan Archives, MCC/CL/L/CC/1/114.

[31] Giles Gilbert Scott, letter to the editor, *Times* (London), 8 May 1940, 9; "Railings for Scrap," *Builder*, 17 May 1940, 581; Gavin Stamp, "Scott, Sir Giles Gilbert (1880–1960)," in *Oxford Dictionary of National Biography* (Oxford: Oxford University Press, 2004).

[32] Note by E. Maplesden, 3 May 1940, NA, WORK 16/1710. Emphasis in original.

[33] "Lincoln's Inn Fields Railings," *Times* (London), 6 Feb. 1941, 2; Helen Jones, "Dalton, (Florence) Ruth, Lady Dalton (1890–1966)," in *Oxford Dictionary of National Biography* (Oxford: Oxford University Press, 2004).

[34] "Scrap Metal from Nelson's Day: Cannon 'Called Up' from the Tower," *Times* (London), 30 May 1940, 4; London County Council, *Minutes of Proceedings, 1940–1942*, 5 July 1940, 234.

East End, Victoria Park lost not only its railings, but also its nineteenth-century bandstand.[35]

In contrast to those who believed that railings should go because they were ugly, old fashioned, or elitist, others maintained that they ought to be preserved because they constituted an important part of British history and culture. Rejecting the government's assertion that most railings were unnecessary, the theater critic Ivor Brown argued that

> to dismiss these protections of private and public property as fussy, Victorian nuisances like antimacassars and draped, overloaded mantle-shelves is nonsensical. . . . It is all very well for the village green to be unfenced, because a village green is in no danger of being trodden to bare earth. But a mid-city park is not a village green. It is a preserve, something saved from brick and stone and the hustling crowd, a shelter and enclave for those who wish to play or to bask. Of this fact the railings are a reminder and, to some extent, a guarantee.

Brown was equally insistent that railings should be maintained around private squares: "If the squares are to be de-fenced and turned into public playgrounds," he declared, "the keyholders have as much right to compensation as if the city council were annexing the lawns behind suburban mansions." In spite of the exigencies of war, Brown insisted that the rights of the individual deserved respect. "What has the Nationalisation of the Means of Production, Distribution, and Exchange to do with granting your neighbours' riotous dogs, and possibly no less turbulent children, the freedom of your flower-beds?"[36] Echoing Brown, the essayist Robert Lynd, writing under the pseudonym Y. Y., declared that even the ugliest of railings performed a civilizing function. "Privacy is obviously one of the great needs of urbanised man, and what better symbol of privacy can he have than railings?" If the Englishman's home was his castle, he insisted, it needed railings just as a castle needed a moat. Connecting the defense of privacy to the defense of the nation, he asserted that railings acted as the homeowner's "English Channel against invasion. They may have lost their value from a utilitarian point of view, but they still retain their poetic significance."[37]

Germany's invasion of France and the Low Countries changed the way British officials looked at railings. As one civil servant noted in early

[35] "Rate of Salvage Quadrupled: 850 Local Authorities Now Taking Part," *Times* (London), 18 May 1940, 11.

[36] Ivor Brown, "A Defence of Fences," *Manchester Guardian*, 4 May 1940, 8.

[37] Y. Y. [Robert Lynd], "Goodbye to Railings," *New Statesman and Nation*, 14 Dec. 1940, 617–18.

June 1940, "As recently as a month ago we understood that what was wanted from the Royal Parks was not so much a weight of metal as a gesture that could be used for publicity. The posts and rail along the East Carriage Drive in Hyde Park (essentially a barrier against crowds) were selected for sacrifice." They weighed approximately thirty tons. Since that time, he observed, "the situation had changed. We were now told that the need for scrap iron was so great that we should throw up all railings we could possibly do without."[38]

Although the government did not possess the legal authority to force the crown to hand over railings that surrounded royal parks, palaces, and other crown properties, government officials feared that leaving these railings in place might provoke populist anger at a time when railings were being removed from many schools, public parks, and government buildings. On 24 May 1940, Lord Claud Hamilton, a close adviser to Queen Mary, telephoned Sir Patrick Duff in the Office of Works to register the queen's displeasure at the prospect of losing the railings outside her residence, Marlborough House. Hamilton added that these railings performed an essential role in preventing unauthorized persons from entering the grounds of the royal residence.[39] In response, the Office of Works decided to leave them standing.[40] In public, of course, government officials said nothing of this dispute. Speaking that month in Glasgow, Judd praised "the tremendously helpful interest shown by Queen Mary in the salvage campaign and the splendid example shown by members of the Royal Household."[41] A short time later the royal family decided to go even further and scrapped the ornamental lampposts that were part of the grounds of Windsor Castle.[42]

Behind the scenes, officials from the Office of Works continued to press the royal family to give up additional railings. In a July 1940 letter to Hamilton, Duff noted,

> We removed with His Majesty's permission the stretch of railings outside Lancaster House and St. James's Palace and, I confess, I had hoped to stop there. But the need for scrap metal is as urgent as ever: as you can see, the Press have been enlisted to

[38] A. A. Rayner to H. E. C. Gatliff, 5 June 1940, NA, T 161/1405.
[39] Note by Patrick Duff on Railings, 24 May 1940, NA, WORK 16/1710.
[40] Claud Hamilton to Patrick Duff, 25 May 1940, NA, WORK 16/1710; Memorandum, 27 May 1940, NA, WORK 16/1710; Memorandum, 29 May 1940, NA, WORK 16/1710.
[41] "The Institute of Public Cleansing: Scottish Centre at Glasgow," *Public Cleansing*, June 1940, 313.
[42] "The King's Scrap Gift," *WTW*, 20 July 1940, 5.

aid the campaign: and the Scrap Metal [i.e., Iron and Steel] Control are clamouring for more and ask why we do not carry on down the Mall. It will be a little embarrassing and pointed to stop short at Marlborough House, and, in the circumstances, would you be ready to agree that we ought to avail ourselves of Her Majesty's permission to take down the railings outside Marlborough House?[43]

The queen agreed, and the Office of Works ordered them removed immediately.[44]

In September 1940, six members of the Chelsea Arts Club registered their opposition to the plans of the Westminster Council "to scrap the noble Georgian railings of Berkeley Square, while there is still an illimitable supply of ugly ironwork of later periods, as in Leicester Square, to draw upon."[45] Despite their pleas, workmen removed these railings in March 1941. A reporter with the *Times* viewed this as an entirely positive development. Using an argument similar to that deployed in the 1960s by advocates of urban "renewal," the journalist wrote that "Berkeley Square was improved by the removal of its iron railings. At the same time much useful scrap was released, and to-day I travelled by train with a useful quota of these railings to an ironworks somewhere in England and there saw them turned into raw material for war purposes." Noting that these railings had survived for more than two centuries, the reporter observed that their destruction took only seconds as enormous factory shears chopped them into short pieces. After being stripped of paint, the broken railings were "heated in furnaces to 1,300 deg. Centigrade. Each box is shot out as a mass of glowing metal and, with a speed born of skill and long practice, is put through a series of operations which sees it emerge as a finished bar of iron anything up to 80ft. long."[46]

In October 1940, the steel industry executive Sir Andrew Duncan assumed leadership of the Ministry of Supply. His predecessors, Leslie Burgin and Herbert Morrison, had managed to stimulate the collection of only 13,000 tons of railings during the first year of the war – almost all of it from parks, schools, and government offices.[47] Duncan was determined to increase dramatically the scrapping of railings.[48] In addition to

[43] Patrick Duff to Claud Hamilton, 23 July 1940 (copy), NA, WORK 16/1710.

[44] Claud Hamilton to Patrick Duff, 29 July 1940, NA, WORK 16/1710.

[45] Philip Connard et al., letter to the editor, *Times* (London), 7 Sept. 1940, 5.

[46] "War Materials from Berkeley Square," *Times* (London), 28 Mar. 1941, 2.

[47] "'Scrap for Victory': Where the Old Iron Goes," *Times* (London), 16 Aug. 1940, 2; "Park Railings: Manchester's Report on Removal," *Municipal Review*, Sept. 1940, 190; "Scrap Iron Helps War – and Rates," *Public Cleansing*, Dec. 1940, 131.

[48] "Scrap the Railings," *WTW*, 9 Nov. 1940, 2.

FIGURE 16. Using brute force to remove railings. Well before the Ministry of Supply ordered the requisition of railings in September 1941, many government departments and some private owners surrendered their railings voluntarily. This photo, taken on 3 March 1941 in Berkeley Square, Mayfair, shows the damage to stonework that often accompanied the removal of railings.
Source: Ministry of Information Second World War Press Agency Print Collection, 1941. Courtesy of Imperial War Museum, © IWM (HU 57684).

taking railings from parks and squares, Duncan launched a campaign to encourage people to donate their house railings to the war effort.[49] One of his allies in this effort was Megan Lloyd George, who chaired the Women MPs' Salvage Advisory Committee. In December 1940, members of the press gathered around her as she used an oxyacetylene torch to remove some of the first household railings donated to the wartime scrap metal campaign.[50] Not to be outdone, Ruth Dalton soon followed suit. In early February 1941, the *Times* published a photo of her, torch in hand, severing "the first section of the 200-year-old iron railings in Lincoln's Inn Fields. It is estimated that these railings will yield about 60 tons of scrap metal."[51]

49 "New Campaign for Scrap," *Times* (London), 4 Nov. 1940, 9; *Parliamentary Debates*, Commons, 30 Jan. 1941, vol. 368, col. 703W.

50 "Miss Megan Lloyd George, M.P., Cuts Railings for Scrap Campaign," *WTW*, 28 Dec. 1940, 8.

51 "Lincoln's Inn Fields Railings: Scrap Metal for War," *Times* (London), 6 Feb. 1941, 2.

A short time later Dalton moved from London to Manchester, where she worked as a liaison officer for the Ministry of Supply.

Duncan's efforts quickly ran into the same legal obstacles that had bedeviled earlier attempts to take private railings. Many people rented the houses or flats in which they lived, and they had no right to dispose of their landlords' property.[52] Homeowners also faced impediments, for often they did not possess freehold to the land on which their houses stood. In such cases, the railings generally belonged to the landlord.[53] In May 1940, a Treasury official had suggested a radical new tactic: "Is it really out of the question," asked J. H. E. Woods, "that the Ministry of Supply should simply make an order giving themselves *compulsory* powers?"[54] This suggestion sparked a flurry of activity as officials examined the legality of taking people's railings without their consent. Colin H. Pearson of the Treasury Solicitor's Department issued a memo in which he stated, "If the use of iron railings as scrap iron is necessary for the prosecution of the war and the maintenance of supplies essential to the life of the community, in my opinion the iron railings can be pulled down and removed." Presciently, he noted that compensation would pose difficult questions.[55] In light of these difficulties, R. D. Fennelly, the official who oversaw scrap iron at the Ministry of Supply, argued against the requisition of private railings. Implementing such an order would be complicated and controversial, and he argued that railings from parks and other government property were providing plenty of scrap for the time being.[56] Ironically, it would later fall to Fennelly to sign the letter that ordered local authorities to requisition railings from private houses.

In the summer of 1940, a Ministry of Supply memorandum noted that "if the collection of scrap from home sources is being pressed energetically, it should be possible to make a substantial improvement on the record of the past year and reduce our dependence upon imported supplies. The requisitioning of scrap, for example, is long overdue, and once air attacks begin on a large scale there will presumably be additional supplies available from damaged buildings and plant."[57] That August, during the height of the Battle of Britain, a member of parliament asked whether the minister of supply would now move to requisition privately

[52] "Railings for Scrap: New Campaign," *Builder*, 8 Nov. 1940, 463.

[53] On the contractual responsibility of tenants to maintain railings, see "Questions in the House," *Public Cleansing*, Mar. 1941, 224.

[54] J. H. E. Woods to R. D. Fennelly, 21 May 1940, NA, T 161/1405. Emphasis in original.

[55] Colin H. Pearson, memorandum, 11 May 1940, NA, AVIA 22/364.

[56] R. D. Fennelly to J. H. E. Woods, 3 June 1940, NA, T 161/1405.

[57] Ministry of Supply, The Steel Position (draft), 10 July 1940, item 22, NA, POWE 5/113.

owned iron railings. Herbert Morrison replied evasively, saying only that he would do so if it became necessary.[58]

BLITZ SCRAP

On 7 September 1940, Germany launched large-scale bombing attacks against London and many other British cities. Over the months and years that followed, aerial attacks killed approximately 60,000 people in the United Kingdom and left behind mountains of destroyed and damaged buildings. The scrap metal that workers salvaged from the rubble became an important source of iron and steel for the manufacture of bombs that British airplanes dropped on German cities. Despite its extensive salvage efforts, Britain faced a growing shortage of iron and steel scrap as its principal overseas supplier, the United States, reduced its exports in support of its own rearmament program. In response, British officials ordered the removal of additional railings from parks and decided to demolish rather than repair a number of moderately damaged buildings simply to recover iron and steel from them.

The onset of the blitz slowed down plans to begin requisitioning railings for two reasons. First, officials were busy fighting fires, clearing the streets, restoring services, making emergency repairs, organizing civil defense, and providing food and shelter to people who had lost their houses. Second, air raid damage to Britain's cities made available thousands of tons of structural steel from destroyed buildings.[59] The man chosen to oversee the clearance of debris in London was the respected civil servant Sir Norman Fenwick Warren Fisher. As special commissioner of the London Civil Defence Region, he supervised a large organization of workers who shored up bomb-damaged structures, demolished others, and transported the detritus to salvage dumps established in Hyde Park, Blackheath, and other sites in London. By November 1940, Warren Fisher (as he was generally known) oversaw a workforce of 9,600, including men from the Royal Engineers and the Pioneer Corps.[60] They salvaged an extensive range of items from bomb-damaged buildings, including roofing slates, window hardware, and electric and gas fittings.[61] These articles, in addition to

[58] "Iron Railings," *WTW*, 31 Aug. 1940, 19.

[59] R. Stanton to R. D. Fennelly, 26 Oct. 1940, NA, HLG 102/92.

[60] "Repairing London Damage: Royal Engineers Show Their Skill," *Times* (London), 19 Nov. 1940, 2; Geoffrey K. Fry, "Fisher, Sir (Norman Fenwick) Warren (1879–1948)," in *Oxford Dictionary of National Biography* (Oxford: Oxford University Press, 2004).

[61] Salvage Operations prior to Demolition, 29 Nov. 1941, NA, HO 186/2005.

steel, bricks, and crushed stone, were used in building repairs, civil defense projects, and military construction. Anything that remained after these needs were fulfilled was sold to scrap merchants.[62] By the time the war ended Fisher's men would remove a third of a million tons of steel, 300,000 tons of wood, 132 million bricks, 2.5 million slates and tiles, 6.7 million tons of hardcore, and nearly 4 million tons of debris from London. Perhaps the most unusual material to be recycled from the rubble of the blitz was the ash from 5,000 tons of tobacco that burned in a warehouse hit by bombs. Workers succeeded in salvaging the nicotine-rich material to be made into insecticide.[63]

The same month that the blitz began, a delegation from the Royal Institute of British Architects (RIBA) asked to meet with the minister of home security to discuss the damage that German attacks were causing to buildings of architectural or historical significance. The matter was handed over to the Ministry of Works, which began to work closely with RIBA. "The great danger," as one official noted,

> is that when buildings of historic importance are damaged, the local authority, either because it does not realise their importance or because it has more than enough other jobs to tackle, will treat them like the other damaged buildings, demolishing the parts that are unstable and clearing up the debris without considering that, by simple measures, walls of possibly irreplaceable interest may either be preserved intact or carefully removed for details of carving or monuments to be sorted out, set aside and stored before the rest of the debris is taken away.[64]

On 29 December 1940, the City of London suffered a devastating raid, which caused enormous loss of life and damaged or destroyed many of its most venerable buildings. St. Paul's Cathedral somehow survived the onslaught, but many of Sir Christopher Wren's other churches did not. In the aftermath of this attack, crews resorted to explosives to bring down many buildings that had been seriously damaged. As the *Times* reported, "Newgate Street was the scene of the first big dynamiting operation. There was a big explosion which caused smoke and dust to float across the City, temporarily obscuring the view of St. Paul's Cathedral."

[62] "Removing London Debris: Salvage of Steel and Masonry," *Times* (London), 26 Oct. 1940, 2.

[63] "The War History of the Architect's Department, 1939–1945" n.d., 54, 58, London Metropolitan Archives, LCC/AR/WAR/1/29.

[64] Patrick Duff to George Gater, 14 Dec. 1940, NA, WORK 14/1180.

Workers also planned to use explosives in Aldersgate, Fore Street, and near London Wall.[65]

The authorities used much of the rubble from this and other raids on London for the construction of military airfields. As an unpublished history of the London County Council's Architect's Department put it, "Filled-in basements were excavated, huge dumps were rapidly cleared; lorries, railway wagons and barges were loaded up and sent . . . to transform fields and meadows into the great runways for aircraft and London reconstructed her ruins into the great take-off for the air offensive to come."[66]

London produced more blitz scrap than any other city in Britain, but other heavily damaged areas also yielded large amounts. Coventry, a city in the West Midlands that was devastated by a bombing raid on 14 November 1940, was one of them. The damage was so extensive that more than a year later thirty-four demolition firms were still at work in the city. Their first priority was to remove steel and other useful materials from damaged buildings.[67] By the summer of 1942 workers had removed 3,000 tons of steel and 115,000 tons of stone, brick, and concrete from the city. Two airfields about ten miles from Coventry took a total of 34,000 tons of crushed stone and brick from the city. Not everything could be recycled, of course. The bombing of Coventry produced 73,000 cubic yards of debris that the authorities disposed of in landfills.[68]

In February 1941, the Ministry of Works began forming panels of architects to "draw up lists of buildings of architectural or historical importance" throughout Britain (except in London, which had its own separate system). After officials at the Ministry of Works approved the

[65] "Demolition Work in the City: Unsafe Buildings Blown up," *Times* (London), 3 Jan. 1941, 4; "The War History of the Architect's Department, 1939–1945" n.d., 49, London Metropolitan Archives, LCC/AR/WAR/1/29.

[66] "The War History of the Architect's Department, 1939–1945" n.d., 54, London Metropolitan Archives, LCC/AR/WAR/1/29; Charles Mendel Kohan, *Works and Buildings*, History of the Second World War, United Kingdom Civil Series (London: HMSO, 1952), 342–3. In October 1940 the Office of Works became the Ministry of Works and Buildings. Less than two years later, its name changed to the Ministry of Works and Planning. Finally, in February 1943, it became known simply as the Ministry of Works (Kohan, *Works and Buildings*, 24). For simplicity's sake I follow Kohan's practice of using the latter term whenever possible.

[67] C. W. Le Grand, minutes of meeting in Coventry, 17 June 1942, NA, HO 207/1169.

[68] City of Coventry, "Clearance work by London County Council, War Debris Survey Organisation: General Report," n.d., NA, HO 207/1169.

selections, they forwarded the lists to local officials. If any listed building were damaged, local authorities would be required to inform the ministry, which would send an architect to visit the building "as soon as possible after rescue or fire-fighting operations have been completed, so that he may advise upon demolition or clearance. After the prior needs of public safety and the restoration of essential services have been considered, he will decide what portions of the building should be shored up, which taken down for preservation, and which are of no importance."[69]

Perhaps the most important building in Britain to sustain severe damage during the war was the Palace of Westminster, more commonly known as the Houses of Parliament. On 10 May 1941, exactly one year after Churchill became prime minister, German bombs obliterated the debating chamber of the House of Commons. The rest of the building would have burned down as well had it not been for the heroic efforts of firefighters. Officials later announced that steel salvaged from the building would be melted down and turned into munitions.[70] This attack marked something of a last gasp for the Luftwaffe; less than two weeks later, major German bombing raids against British cities came to an end. Although most people in the United Kingdom welcomed the end of the eight-month period known as the blitz, the officials responsible for raw materials viewed it as something of a mixed blessing because it halted the "production" of blitz scrap. As a member of the Harriman Mission later explained in a report to Washington, "Lack of bombing has naturally decreased the supply of iron and steel scrap from damaged buildings and it is worthy of comment that the scrap derived from this source has been very useful."[71] In 1941 alone Sir Warren Fisher's organization recovered 200,000 tons of steel from bombed buildings in London.[72] These figures, though extremely large, include only the steel that was removed from

[69] Ministry of Home Security, Air Raid Precautions Department, Home Security Circular No. 44/1941, 15 Feb. 1941, NA, WORK 14/1180. The London County Council had its own War Debris Survey office, which supervised panel architects in the London region and directed repairs there. See "London Region, Out-Counties Groups 6, 7, 8, and 9, Buildings of Historic and Architectural Interest: Panel of Architects Scheme," Feb. 1942, NA, WORK 14/1180. For a wartime account of the National Buildings Record, see B. H. St. J. O'Neil, "Historic Buildings and Enemy Action in England," *Journal of the Royal Institute of British Architects*, Mar. 1945, 132–3.

[70] "The House of Commons," *Builder*, 6 Feb. 1942, 120.

[71] M. A. Colebrook, "British Iron and Steel Scrap Collections: September, 1939–December 31, 1941," 2 Mar. 1942, NARA, Record Group 169, A 1 500, Box 1670, File 412841.

[72] Notes of a conference at the Ministry of Works, 5 Jan. 1942, NA, HO 207/76.

buildings that were damaged so severely that they were deemed to be beyond repair. Additional steel came from buildings that were not as badly damaged.

In July 1941, the Ministry of Home Security informed local councils across Britain that there was an "urgent need for utilising all such iron and steel in the quickest and most efficient manner for the furtherance of the national war effort." As a result, some buildings that were not damaged beyond repair and did not constitute a threat to public safety would nonetheless be demolished for the sole purpose of recovering the iron or steel that they contained.[73] The suggestion that large numbers of damaged buildings should be scrapped rather than repaired proved highly contentious. Five days after the Pearl Harbor attack, a conference took place at the Home Office between officials from the Ministry of Home Security, the London Civil Defence Region, the Ministry of Works, and the Ministry of Supply. Major Holt, an official from the Ministry of Supply, explained that he and his colleagues

> were very much concerned with difficulties which had arisen in connection with the extraction of iron and steel from bombed buildings up and down the country owing to the intervention of the War Damage Commission. The matter had come to a head owing to the case of fourteen buildings in the City of London which the representatives of the War Damage Commission had refused to admit to be proper subjects for demolition, at any rate without a prolonged and careful examination. Similar trouble had arisen in Manchester and elsewhere.[74]

Britain faced an acute shortage of steel, Holt argued, and the U.S. decision to suspend exports of scrap metal to Britain meant that the situation was "much too urgent" to allow a delay of even a few weeks. A. B. Valentine of the Ministry of Home Security, who chaired the meeting, suggested that the best way to deal with the meddling War Damage Commission would be to persuade it that a building was unsafe. Curiously, Valentine did not explain how ministry officials were to convince local engineers to condemn buildings that the War Damage Commission thought should be preserved. The fundamental issue, observed the director of salvage and recovery, G. B. Hutchings, was a concern on the part of the commissioners "that a certain

[73] Ministry of Home Security, Air Raid Precautions Department, Home Security Circular No. 156/1941, 16 July 1941, Trades Union Congress Collection, Modern Records Centre, Warwick University, MSS. 292/557.61/1. According to this document, the policy of sacrificing bomb-damaged buildings so that materials could be harvested from them had been announced some time earlier in Home Security Circular No. 127/1941, but I have been unable to locate a copy of this circular.

[74] Notes of a meeting at the Home Office, 12 Dec. 1941, NA, HO 207/76.

partial loss [damage from enemy attack] was being turned into a greater one by the extraction of the steel in question."[75]

In April 1942, an official with the London Civil Defence Region issued a memorandum on the collection and disposal of scrap metal. In it, he defined blitz scrap as "iron and steel from buildings damaged beyond repair by enemy action" and declared that "all other scrap including that from repairable war-damaged buildings will be termed 'National Survey' Scrap." In the case of the latter, he noted that the regional salvage officer first had to supply the group surveyor or group engineer with a "pink form authorising demolition and giving specific instructions for collection and despatch."[76]

This practice continued until March 1943, when the Ministry of Works issued a detailed set of new procedures for dealing with steel from repairable buildings, which made it clear that the removal of iron and steel should not occur unless the damage that this entailed did "not greatly increase the cost of the ultimate repair." In addition, it specified that Sir Warren Fisher and his colleagues would discuss the proposed work with both the property owners and the War Damage Commission. If "serious objections" came from either of them, the dispute would be referred to the Ministry of Works.[77] It seems remarkable that such steps were not taken prior to this point.

As the Allied air campaign against Germany intensified, the Warren Fisher organization faced an enormous demand from the military to supply hardcore (crushed bricks, stone, and concrete) for the construction of airfields.[78] In a November 1942 letter to Hugh Beaver, director general of works and buildings, one of Fisher's staff wrote,

> We are hoping that it may be possible to avoid pulling down more buildings than is necessary for the purposes of safety, etc. Fortunately we have one or two dumps with about half a million cubic yards of hardcore in them, and we are drawing on this source to secure the best deliveries of hardcore with the available labour.
>
> Nevertheless it is only fair to warn you that the demands which we are receiving from the Air Ministry are going up by leaps and bounds.[79]

[75] Notes of a meeting at the Home Office, 12 Dec. 1941, NA, HO 207/76.
[76] London Civil Defence Region, Memorandum No. 18, 30 Apr. 1942, NA, AO 30/32.
[77] London Civil Defence Region, Memorandum No. 27, 3 Mar. 1943, NA, HO 207/76.
[78] Norman Fenwick Warren Fisher to J. H. Burrell, 28 Nov. 1942, NA, HO 186/2005.
[79] A. J. Shove to Hugh Beaver, 27 Nov. 1942, NA, HO 186/2005.

In addition to using hardcore to build runways, in December 1943 the government directed 2,600 cubic yards of the material per day to the manufacture of Mulberry harbors (massive concrete artificial harbors used in the D-Day invasion).[80] So great was the demand for crushed stone that Fisher's organization considered retrieving rubble that workers had used to fill in sand pits and wetlands earlier in the war. Officials also looked to "slum clearance" as a source of hardcore when bombing raids were failing to produce sufficient quantities of rubble.[81]

[80] London Civil Defence Region, meeting notes, 15 Dec. 1943, NA, HO 186/2005.

[81] London Civil Defence Region, meeting notes, 16 Feb. 1944 and 15 Mar. 1944, NA, HO 186/2005.

8

Wasting Paper

> Ink preserves the past; paper must secure the future. . . . Muniments are asked to become munitions, and millions of written and printed pages must now be divided into matter and spirit, so much material and such and such black marks made upon it.
>
> –*The Times*, 1941[1]

When people think of important strategic materials, paper does not generally come to mind. Yet paper was a major ingredient in such varied products as propaganda posters, bullet cartridges, shipping containers, land mines, and radio components.[2] In wartime, the manufacture of armaments and explosives in Britain consumed more than twice as much paper as the production of books and magazines.[3] The now-ubiquitous icon for recycling implies that materials can be used again and again in a closed loop. Paper cannot, however, be recycled an infinite number of times, for each time it is processed, its fibers become shorter. And paper incorporated into munitions is of course destroyed when they explode. For this reason, even the most "efficient" use of resources in wartime

[1] "Paper against Ink," *Times* (London), 14 Nov. 1941, 5.

[2] "Waste Paper Needed for Munitions: Lord Beaverbrook Asks for 100,000 Tons," *Times* (London), 22 Oct. 1941, 4; "Waste Paper for Munitions: Salvage Contest Welcomed," *Times* (London), 5 Jan. 1942, 2; Ministry of Supply, *Salvage: Lectures to Schools and Test Papers (With Answers)* (1942), 2–3, copy in NA, AVIA 22/3088. For an earlier analysis of some of these issues, see Peter Thorsheim, "Salvage and Destruction: The Recycling of Books and Manuscripts in Great Britain during the Second World War," *Contemporary European History* 22, no. 3 (2013): 431–52.

[3] "Comparative Consumption of Paper and Board in the United Kingdom . . . " [1942], NA, BT 28/1085.

results unavoidably in the permanent loss of raw materials. Recognizing these realities, an American official who was based in London commented in January 1942 that at least half of the paper consumed by British munitions factories "either leaves the country or is rendered useless."[4] As the war progressed, Britain's paper mills produced ever lower grades of paper, which were less suitable for repulping than the longer fiber papers produced before the war. As one British expert explained in August 1943, "The supply of waste paper must tend to dwindle with the reduced amount of paper flowing into consumption and the quality is deteriorating markedly as some waste paper is now doing its second or third journey round. The injection of fresh virgin fiber into the paper making system must therefore be kept in mind."[5]

As government and industry pressed ever harder to supply pulp mills with high-quality sources of fiber, some warned that the indiscriminate salvage of books and documents was adding to the Nazis' destruction of history and culture. Little more than a fortnight after Germany invaded Poland, Wilfrid Greene, president of the British Records Association, charged that paper drives during what he referred to as "the last War" had destroyed "many records of great importance for the study of social, industrial, and political history. Accumulations of old papers . . . were handed over all the more readily because they were not in current use. But it was precisely among papers of this kind that valuable material was hidden and the loss to historical research occasioned by its destruction is deplorable."

Anxious to prevent similar self-inflicted damage from recurring, Greene and others worked hard between 1939 and 1945 to prevent important documents from being sent to paper mills. To help owners decide which of their papers should be spared from wartime salvage collections, Greene offered the free assistance of his organization. He promised that doing so would not have a detrimental impact on efforts to stimulate paper recycling, for he argued that only a small fraction of documents was worth saving.[6]

[4] M. A. Colebrook, "War-time Salvage of Household Waste in the United Kingdom," 26 Jan. 1942, NARA, Record Group 169, A 1 500, Box 1816, File 456826.

[5] United Kingdom Paper in Wartime, 23 Aug. 1943, 2, NA, BT 28/1085.

[6] Wilfrid Greene, letter to the editor, *Times* (London), 21 Sept. 1939, 9. For a summary of Greene's life and career, see Jeremy Lever, "Greene, Wilfrid Arthur, Baron Greene (1883–1952)," in *Oxford Dictionary of National Biography* (Oxford: Oxford University Press, 2004).

Another dedicated advocate for the preservation of old papers in wartime Britain was Joan Wake, who led the Northamptonshire Record Association from 1920 to 1963. Nearly a year before the outbreak of war, she had warned that efforts to clear attics of flammable material might erase "centuries of evidence sorely needed by the student of our local, social, legal, and economic history."[7] Her concerns grew stronger as the war progressed. In November 1940, *Waste Trade World and the Iron and Steel Scrap Review* reported, "Miss Joan Wake declared at the meeting of the British Records Association in London that we were destroying historical records with more than Teutonic thoroughness. 'If English history does not matter, all this destruction does not matter in the least,' she said, 'and the sooner we boil down the Domesday Book to make glue for aeroplanes the better.'"[8]

In June 1941, prompted by concerns about the loss of records from the twin dangers of enemy action and thoughtless salvage, members of the Historical Manuscripts Commission and the British Records Association began to compile a centralized record of the existence and location of manuscript collections throughout Britain.[9] This invaluable resource, which came to be known as the National Register of Archives, is now integrated into Discovery, the online catalog of Britain's National Archives.[10] A short time later, W. C. Berwick Sayers, chairman of the Library Association, charged that the government's overly zealous salvage efforts threatened not only unpublished documents, but also books. Sayers worried that citizens might, in their enthusiasm to contribute to the war effort, unknowingly hand over for pulping valuable volumes that ought to be saved. "Unless there is discrimination, a famine in copies of quite important books may result."[11]

Unfortunately, the warnings of preservationists had little effect. Most people insisted that the important thing was to win the war – by any means necessary. The pulping of unpublished documents, as well as rare books and periodicals, added immeasurably to the losses caused by enemy bombs. Although far fewer tons of bombs fell on the United Kingdom than on Germany, British libraries nonetheless experienced enormous destruction.

[7] Joan Wake, letter to the editor, *Times* (London), 24 Jan. 1939, 13.
[8] "Ancient Records for Pulp?" *Waste Trade World and the Iron and Steel Scrap Review* (hereafter *WTW*), 23 Nov. 1940, 15.
[9] C. T. Flower, letter to the editor, *Times* (London), 5 June 1941, 5.
[10] National Archives of England, Wales, and the United Kingdom, Discovery catalog, discovery.nationalarchives.gov.uk.
[11] W. C. Berwick Sayers, letter to the editor, *Times* (London), 26 Sept. 1941, 5.

The National Central Library, which served as a clearinghouse and reservoir for interlibrary loan in Britain, suffered severe damage from an air raid on the night of 16–17 April 1941, when fire and water destroyed 104,349 volumes.[12] The damage would have been far worse had much of the library's collection not been evacuated from London to Hemel Hempstead the previous year.[13] Less than a month after the National Central Library fire, incendiary bombs fell on the British Museum Library in Bloomsbury. The ensuing fire turned the southwestern part of the building into a virtual blast furnace, incinerating 125,000 books.[14] Additional attacks damaged or destroyed many other academic and specialist libraries during the war years, including those of Birkbeck College, Goldsmiths' College, University College London, the Birmingham Natural History and Philosophical Society, and the Manchester Literary and Philosophical Society. The residents of Coventry, Exeter, Hampstead, Liverpool, Plymouth, and Richmond saw their public libraries go up in flames.[15]

The task of rebuilding these collections proved difficult for many reasons. The magnitude of war-related losses left many libraries in search of an entire replacement stock of books, but the supply of both new and antiquarian editions was severely affected by the conflict. Bombs had destroyed not only libraries but also printing presses, type, and publishers' warehouses, and the supply of paper for new books remained scarce long after the war ended. These problems were compounded by wartime salvage drives. Walter Harrap, whose father George had founded the publishing house that bore the family name, noted that the national need for scrap metal had "resulted in publishers sacrificing a high percentage of their plates, blocks, and standing type from which reprints could have been made when increased supplies of material became available."[16]

12 The National Central Library, *26th and 27th Annual Report of the Executive Committee, 1941–42 and 1942–43* (London: n.p., 1943), 7–8; "The National Central Library Fire," *Library Association Record* 43 (May 1941): 88.

13 For an account of the origins and evolution of this library, see Reginald Harrison Hill, "The National Central Library: Impressions and Prospects," *Library Association Record* 47 (Dec. 1945): 246–52.

14 British Library Corporate Archives, Minutes, Reports, Letters, etc., 12 July 1940, British Library, London, DH 2/106, section III, 59; John Forsdyke, "The Museum in War-Time," *British Museum Quarterly* 15 (1941–50): 1–9, esp. 8.

15 The National Central Library, *26th and 27th Annual Report of the Executive Committee, 1941–42 and 1942–43* (London: n.p., 1943), 9.

16 Walter G. Harrap, letter to the editor, *Times* (London), 17 Oct. 1942, 5. On the family and the firm, see Adrian Room, "Harrap, George Godfrey (1868–1938)," in *Oxford Dictionary of National Biography* (Oxford: Oxford University Press, 2004).

By the autumn of 1941, one year after Germany and Britain began to attack each other's cities, bombing had consumed an estimated twenty million books in Britain and an untold number of unpublished documents.[17] Although the Public Record Office in Chancery Lane emerged from the war relatively unscathed, bombs and rockets destroyed vast numbers of irreplaceable documents held at the War Office, at the Treasury, and in numerous local archives, businesses, and homes across Britain.[18]

Large numbers of books, as well as business correspondence, personal papers, and government records, disappeared in British paper drives during the war. In 1943 alone, Britons contributed 600 million books for recycling – thirty times as many volumes as the Luftwaffe destroyed during its most intensive year of raids against Britain.[19] Praising British efforts to salvage books, one observer quipped, "Hitler once held a great book-burning fiesta in the Reich. It was an extravagant and futile effort of Nazi savages to extinguish the culture of Europe. To-day we are also destroying books, but they are being turned into munitions to blast Hitler out of occupied Europe."[20] In language eerily similar to that which accompanied the burning of books in Nazi Germany, some British moralists suggested that the wartime need for waste paper created a welcome opportunity for "a thorough purge of the civic libraries, long overdue." This would remove "large numbers of trashy, not to say nasty, novels" and allow libraries to "serve some really useful purpose, instead of corrupting the minds of youthful readers."[21]

On 7 September 1941, the first anniversary of the start of the blitz, the Princess Royal (the future Queen Elizabeth II) spoke over the BBC on the subject of salvage. She commended the people of Britain for having contributed 400,000 tons of waste paper to the war effort, but she explained that much more was needed. In addition to making sure that they did not throw waste paper in the refuse bin, she urged everyone to go through their homes to locate other sources of paper that could be given to the nation. "Tucked away in our bookshelves and cupboards and

[17] Ethel Wigmore, "The War and British Medical Libraries," *Bulletin of the Medical Library Association* 34, no. 3 (1946): 151–66, esp. 151.

[18] C. T. Flower, "Manuscripts and the War," *Transactions of the Royal Historical Society*, 4th ser., 25 (1943): 15–33, esp. 21–2.

[19] W. C. Berwick Sayers, "Britain's Libraries and the War," *Library Quarterly* 14, no. 2 (1944): 95–9. Sayers estimated that 95 percent of these books were pulped and the remainder distributed to war-damaged libraries and members of the armed forces.

[20] "Books Wanted," *Public Cleansing and Salvage*, Feb. 1943, 195.

[21] C. Tearup, letter to the editor, *Bath Weekly Chronicle and Herald*, 17 Jan. 1942, 14.

offices there are masses of books, magazines, old business records, and papers of every sort laid by in case they may be wanted some day. That day has come! Your country wants them now. Keep your family Bible of course and anything of historical or special value; but from the rest you can each contribute your special share of the ship loads that these old books and paper would save."[22] Local government officials, business firms, and millions of ordinary citizens responded enthusiastically to this call for waste paper.

In response to the government's message that no amount of paper was too small to salvage, one reader of the *Times* suggested cutting out and recycling the blank pages found at the beginning and end of books. Another urged people to donate the paper that lined their cupboards and chests of drawers.[23] Instead of encouraging such inconsequential efforts, the government focused on a much richer source of salvage: entire books and accumulations of documents. Unfortunately, officials in the Ministry of Supply made no provision for important books and manuscripts to be saved from pulping when they launched these waste paper drives. Many librarians, archivists, and scholars feared that important and irreplaceable elements of Britain's history and culture were "being patriotically destroyed."[24]

At the same time the government sought to uncover sources of recyclable paper in people's attics and bookcases, it looked to public agencies as another rich source. On 1 July 1940, officials from the Ministry of Health, in wording that was identical to an announcement that the Local Government Board had issued in 1918, called on local authorities to recognize the "paramount importance of releasing for repulping papers and books which it is no longer necessary to retain." The directive urged officials to preserve documents that "possess historical importance," were still in use, or which might be needed for administrative or judicial purposes in the future. Everything else was to be donated to the war effort.[25] Referring to "the vast accumulation of ancient files of papers" held by government departments, Sir William Davison, MP, suggested in July 1941 that the wartime need for waste paper provided a perfect

[22] "Salvage from the Home: Princess Royal's Appeal," *Times* (London), 8 Sept. 1941, 2.
[23] "Waste Paper for Munitions: Salvage Contest Welcomed," *Times* (London), 5 Jan. 1942, 2.
[24] Carl M. White and P. S. J. Welsford, "The Inter-Allied Book Centre in London," *Library Quarterly* 16, no. 1 (1946): 57–62, quotation on 58.
[25] Ministry of Health, "Circular 2073: Waste Paper," 1 July 1940, NA, HLG 102/91. For comparison, see Local Government Board, "Waste Paper," 30 Aug. 1918, NA, HLG 102/93.

chance "to get rid of all this valueless clutter."[26] Officials soon dispatched to the mills a large collection of historical documents, including a significant number of financial records created by the East India Company and the Board of Control.[27]

When officials in Aberdeen voted to recycle all the correspondence files of their education department that were more than a decade old, they specifically "decided to ignore any historical or other interest which these letters might have."[28] Surrey County Council went even further in proposing to pulp most of its papers, books, and documents that were more than three years old, including 150,000 files on persons who had received public assistance. The council estimated that the disposal of these files would contribute fifteen to twenty tons of paper for the war effort and free up space in the county hall.[29] Six months later a coalition of local government bodies called on officials throughout the United Kingdom to recycle virtually all case files older than 1930 related to poverty relief. Although its letter recommended that "specimens" be preserved for posterity, it implied that very few such records needed to be retained.[30]

Most businesses similarly welcomed the government's exhortation to donate paper for the war effort, as it relieved them of the long-established expectation, in some cases required by law, that they store the voluminous records of their past activities for many years. "Legal firms," noted one observer, "have found particularly rich stores of bygone correspondence and documents that are now out of date."[31] Taking the salvage campaign to heart, in January 1941 the *Times* cleared nineteen tons of ledgers and other business records from its own offices.[32] Less than a week later, the Aberdeen *Press and Journal* boasted that "all the old letter books, letter files, day books and ledgers, and many old reference books belonging to the firm are being handed over to the paper mills. For years they have been lying in piles in a cellar. Now they will go to make new

[26] *Parliamentary Debates*, Commons, 3 July 1941, vol. 372, col. 1470; "The Trade in Parliament," *WTW*, 12 July 1941, 6.

[27] Antonia Moon, "Destroying Records, Keeping Records: Some Practices at the East India Company and at the India Office," *Archives* 119 (2008): 114–25, esp. 119.

[28] "Voted for Scrap: Old Letter Books," *Press and Journal* (Aberdeen) 13 June 1942, 3.

[29] Public Assistance Department of Surrey County Council to the Ministry of Health, 26 Nov. 1941, NA, HLG 102/91.

[30] County Councils Association and the Association of Municipal Corporations, circular letter, 26 May 1942, Walsall Local History Centre, 235/3.

[31] "More Waste Paper Being Salved," *Times* (London), 28 Oct. 1941, 2.

[32] "Schemes for Paper Economy," *Times* (London), 14 Nov. 1941, 2.

paper and . . . the many items of munitions for which paper is so necessary. They weigh fully two tons."[33]

In September 1941, the publicity department of the Ministry of Supply gave local authorities a sample letter to send to businesses within their jurisdiction. This letter, which the ministry hoped would appear to originate from the local council rather than from Whitehall, was to "be duplicated (if the number is too large to permit individual typing) on your Official paper and signed by the Lord Mayor, Mayor, Chairman of the Council or Salvage Officer."[34] In contrast to the large amount of salvage publicity that targeted women, this letter called on "every business and professional man to play his part . . . by carefully combing his offices and handing over . . . every scrap of paper for which he has not a pressing, immediate use and I do most particularly ask for old ledgers, old correspondence, old books and old documents of every kind." The letter said nothing about preserving records of historical significance. On the contrary, it added that "even though your papers may be of a confidential nature, you need have no hesitation whatever in parting with them. You have my assurance that, if packed in sacks, parcels or cartons, they will be despatched immediately to the mills for instant re-pulping."[35]

In a similar attempt to allay fears about handing over sensitive papers for salvage, the Waste Paper Recovery Association employed an illiterate man to process papers from law firms. The press soon reported that "the non-reader got busy and collected more than five tons of confidential documents and ledgers" from a single office, packing up "century-old papers, old statute books and legal testimonies."[36] The most promising approach, in the view of many recycling proponents, was to use a novel technology, the paper shredder, to eliminate the information content of papers destined for recycling.[37] Unfortunately, these efforts to ensure privacy meant that no opportunity existed save historically significant documents from destruction.

In January 1942, the *Times* reported that police officers in Northamptonshire had aided the national drive for waste paper by

33 "Two Tons of Paper," *Press and Journal* (Aberdeen), 6 Jan. 1942, 3.

34 Benet Williams, circular letter to local authorities in Lancashire, 16 Sept. 1941, Ulverston Urban District Council Records, Cumbria Records Office, Barrow, BSUD/U/S/Box 10.

35 Ministry of Supply, Draft Letter for Use by Local Authorities, 1 Oct. 1941 [postdated], Ulverston Urban District Council Records, Cumbria Records Office, Barrow, BSUD/U/S/Box 10.

36 "Man Who Cannot Read Is Now Salvage Sleuth," *Public Cleansing*, Jan. 1942, 146.

37 "'Targets' for Paper Salvage: Street Stewards Needed," *Times* (London), 3 Mar. 1942, 2.

clearing out "over a ton of old crime records and dossiers." The same article praised a boy in Plymouth for having "nobly surrendered his collection of cigarette cards" and commended musicians for donating their old scores.[38] Not to be outdone, King George VI contributed more than a ton of "waste paper," which consisted of "a large consignment of old books and manuscripts from the royal library."[39] A month later the king's mother, Queen Mary, announced that she would donate papers from her residence, Marlborough House. Following the official line that emphasized the direct connection between paper salvage and munitions, the press reported that her "old letters, records, and files . . . are to be made into cartridge cases."[40] In Sheffield, one woman handed in some five hundred letters that her son had written to her during his fifteen years as a missionary in Canada. As *Public Cleansing*, the leading organ of the waste and scrap trades, reported, "It was always intended that they would make interesting reading when he came home. But in the national emergency Mrs. Jacklin has decided to send them for paper salvage."[41]

Praising the efforts of local authorities to recycle paper, the controller of salvage, Harold Judd, announced in September 1941 that they were "saving the equivalent of 250 ship voyages a year."[42] Just a month later, however, Lord Beaverbrook, who had taken the helm of the Ministry of Supply in June, suggested that not nearly enough was being done on the paper front. Always eager to one-up his colleagues, Beaverbrook applied the same methods to the drive for paper that he had used to call for aluminum when he had led the Ministry of Aircraft Production. On 21 October 1941, he declared, "The Ministry of Supply needs forthwith 100,000 tons of waste paper. This waste paper will be used to make shells and cartridge cases. It is vital to the production of munitions." He added that having this waste paper would reduce reliance on paper-making supplies from North America and free up additional ships to carry munitions to Britain's new ally, the Soviet Union.[43]

Lending its support to the new drive, the *Times* asserted that by donating unwanted books and papers, "The citizens of these islands can

[38] "Salving the Waste Paper: Music Scores Wanted," *Times* (London), 27 Jan. 1942, 2.
[39] "Salvage at Windsor: Ton of Paper Sent from the Castle," *Times* (London), 19 Jan. 1942, 2.
[40] "Queen Mary Helps Salvage," *Gloucestershire Echo*, 5 Feb. 1942, 3.
[41] "Son's 500 Letters as Salvage," *Public Cleansing*, Mar. 1942, 194.
[42] "Saving Waste: Paper, Cinders, and Scrap Metal," *Times* (London), 18 Sept. 1941, 9.
[43] "Waste Paper Needed for Munitions: Lord Beaverbrook Asks for 100,000 Tons," *Times* (London), 22 Oct. 1941, 4.

FIGURE 17. Turning paper into pulp. This wartime photograph was taken at the Thames Board Mills complex in Purfleet, Essex, purportedly the largest producer of cardboard in the world.
Courtesy of Westminster City Archives, WccAcc 152/5/10/15a.

immensely reinforce the firing power of the Royal Navy, Army, and the Royal Air Force, and the Armed Forces of our Allies." Paper, it added, could even be used instead of steel for the manufacture of petrol containers.[44] Echoing Beaverbrook's warning that the nation faced a critical shortage of waste paper was the firm Thames Board Mills, which operated two of the largest paper recycling plants in Europe. The previous year this company had issued a seven-and-a-half minute documentary, *Raw Material Is War Material*, in an effort to convince viewers of the importance of recycling to munitions production.[45] Now, its director once again called on the public, announcing, "We have been reduced to

[44] "Save Unwanted Paper: Vital Needs of War Factories," *Times* (London), 23 Oct. 1941, 2.

[45] "Islington in the Forefront," *Public Cleansing*, June 1940, 335; Thames Board Mills to Cleansing Superintendent, Walsall, 11 Jan. 1940 (copy), Walsall Local History Centre, 235/1.

working at half capacity on account of the shortage of waste paper, and Government orders are being delayed in consequence."[46]

Shortly after Beaverbrook launched his paper appeal, the author Vera Brittain wrote to the *Times* "to deplore the present campaign for the indiscriminate destruction of documents and records in the interests of salvage. While it is, of course, important to writers and publishers, as well as to the direct war interests, that every scrap of paper should be retrieved, it is even more important to avoid exposing British culture to forms of vandalism comparable to those which have destroyed the culture of Central Europe. This threat is especially alarming," she added, "when so many valuable books and documents have already been lost through air raids." Brittain reserved her strongest criticism for calls to destroy the underappreciated sources that recorded the nation's history, such as "ledgers, diaries, note-books, and letters." The more recent the document, the greater the danger that people would not recognize the importance of saving it, she warned.[47] Brittain was far from the only observer to criticize the extremes of wartime paper salvage. In 1942, a columnist for the *Times* warned that "in this matter of throwing away there lurks even in the mildest of us something positively tigerish." Once people began to discard their old papers, the author suggested, they quickly abandoned all restraint. "When we decide reluctantly that some letters must go and begin to tear them across, the rending sound is a savage music in our ears. We have metaphorically tasted blood and are soon red to the elbows." As a result of this lust for destruction, many invaluable documents were being destroyed.[48]

PROMOTING SALVAGE

By 1942, Britain's newspapers had been forced to reduce their consumption of newsprint to just 20 percent of prewar levels. To stimulate greater public interest in recycling, the publishing industry offered prizes to the communities that were able to collect the largest quantities of paper and books for the war effort. The *Times* soon reported that "in many towns where great efforts are being made to secure a prize in the £20,000 competition a bundle of waste paper has become the ticket of admission to cinemas, football

[46] "Saving Waste Paper: Good Results of New Campaign," *Times* (London), 30 Oct. 1941, 2.

[47] Vera Brittain, letter to the editor, *Times* (London), 30 Oct. 1941, 5; Alan Bishop, "Brittain, Vera Mary (1893–1970)," in *Oxford Dictionary of National Biography* (Oxford: Oxford University Press, 2004).

[48] "Old Letters," *Times* (London), 5 Feb. 1942, 5.

FIGURE 18. Women sorting waste paper. In 1943 these women in Suffolk took part in the laborious process of preparing paper for the pulp mills. After removing "contraries" such as string and paper clips, they sorted it into grades based on color, weight, and cleanliness.
Source: Ministry of Information Second World War Official Collection. Courtesy of Imperial War Museum, © IWM (D 14074).

matches, and other pastimes."[49] The month-long contest proved a resounding success from the organizers' perspective. The amount of paper collected by local authorities in January 1942 reached 100,000 tons, 50 percent more than in any previous month of the war.[50] An added benefit of contests, in the eyes of officials in the Ministry of Supply, was that they discouraged people from giving or selling materials to rag-and-bone men. As one salvage official from Manchester noted in 1943, "only collections made by Local Authorities and collections by Voluntary Organisations handed over for disposal by Local Authorities have been included."[51]

49 "Salving the Waste Paper: Music Scores Wanted," *Times* (London), 27 Jan. 1942, 2.
50 "John Gilpin's Deal in Land: Deed Found in Waste Paper at Olney," *Times* (London), 17 Feb. 1942, 2.
51 Ministry of Supply, Honorary District Adviser's Office, Manchester, "£1,000 Book and Waste Paper Drive 'Lancashire,' 1st to 30th September 1943," Ulverston Urban District Council Records, Cumbria Records Office, Barrow, BSUD/U/S/Box 10.

Local officials across Britain went to extraordinary lengths to root out books and paper for the pulp mills. During a drive in December 1942, officials in Birmingham launched a massive publicity campaign, which included a letter from the lord mayor to every schoolchild, and special notices in each of the city's 3,500 buses and trams. All this work paid off. Over a two-week period, the people of Birmingham contributed nearly three-quarters of a million books to the war effort.[52] Coventry's drive was less successful, perhaps because it had lost so many books in the heavy bombing raid that destroyed much of the city in November 1940. Over the same fortnight, its drive brought in 51,477 books, barely a fifth of the target that officials had set for the city.[53]

To encourage people to hand over their books, the Ministry of Supply established numerical goals for each community. Unconsciously mimicking the bombastic and often unrealistic Soviet appeals for workers to exceed their production quotas, in March 1943, the *Times* reported that in "many cases the target-figure is enormously exceeded. Canterbury more than trebled its target of 25,000. Llanfairfechan Rural District, which set out to collect 6,500 books, has more than quadrupled this figure, and Bangor, aiming at 30,000, has reached the astonishing result of 156,111 – more than five times its object." To promote this book drive, the king contributed a thousand of his own books, and Princess Elizabeth and Princess Margaret donated volumes from their collections.[54]

Many people considered recycling contests inherently unfair because communities that had worked hardest in the past would have fewer stocks of recyclable materials remaining. Contests also assumed that communities of comparable size were capable of salvaging the same quantity of metal, paper, and kitchen waste, yet a strong correlation existed between affluence, consumption, and the generation of waste.[55] Well-off city residents, for example, often subscribed to multiple magazines and daily newspapers, and they purchased large numbers of goods wrapped in paper and cardboard. Working-class urban dwellers, as well as rural residents of all income levels, used considerably smaller quantities of

[52] W. H. Andrews, report on book drive in Birmingham in Dec. 1942, n.d., Shakespeare Centre Library and Archive, Stratford-upon-Avon, BRR 55/14/31/4/2.

[53] E. H. Ford, report on book drive in Coventry in Dec. 1942, n.d., Shakespeare Centre Library and Archive, Stratford-upon-Avon, BRR 55/14/31/4/2.

[54] "Royal Gifts of Books," *Sunday Times* (London), 7 Mar. 1943, 7.

[55] John C. Jorden, "Salvage in the Midlands," n.d. [1946], 5, Shakespeare Centre Library and Archive, Stratford-upon-Avon, BRR 55/14/31/4/3.

paper. Criticizing the absurdity of expecting everyone to contribute an equal amount of waste paper, one salvage official, A. W. Ward, argued that the only way for some individuals to satisfy the government's salvage targets would be to increase their paper consumption – clearly something that would be counterproductive to the war effort.[56]

To promote salvage, the government relied not only on exhibitions and contests but also on posters, newspaper advertisements, and pamphlets. In December 1942, the Ministry of Supply, in conjunction with the Salvage and Recovery Board, helped to produce a short book called *The Amazing Story of Weapons Made from Waste*. Its purpose was unabashedly propagandistic. A cover letter that its publisher, Odhams Press, sent to local councils along with an advance copy of the book expressed the hope that "this fascinating story of how salvaged materials are converted into weapons of war will obtain a wide circulation and further stimulate public activity in the salvage of paper and other materials needed for the war effort." Odhams went so far as to include a "suggested draft" review of this book.[57] The publication of *The Amazing Story of Weapons Made from Waste* may well have been influenced by Beaverbrook, who was an old friend of the firm's managing director, J. S. Elias.[58]

By April 1942, so many books were being handed in for recycling that efforts to collect reading material for members of the armed forces were suffering.[59] Salvage mania also had other consequences, as the bookseller W. A. Foyle explained in a letter to the *Times* a month later:

> Many priceless, rare, and irreplaceable books are being destroyed owing to the campaign for waste. Everyone, of course, wants to assist in giving waste paper and books to the Government, but there should be a sifting process to prevent many of our finest books being thrown away. Recently a perfect copy of von Gerning's "Tour along the Rhine," with colour plates by Ackermann was sent in to us as salvage together with other fine volumes. Luckily we can stop their going to the mills, but all over the country these books are being pulped.[60]

[56] Ministry of Supply, Salvage Department, District No. 3, notes of a meeting in Birmingham, 1 Oct. 1941, Shakespeare Centre Library and Archive, Stratford-upon-Avon, BRR 55/14/31/4/1.

[57] *The Amazing Story of Weapons Made from Waste* (London: Odhams, 1942); C. L. Shard, circular letter to local authorities, 7 Dec. 1942, Shakespeare Centre Library and Archive, Stratford-upon-Avon, BRR 55/14/31/3/5.

[58] See Lord Beaverbrook to J. S. Elias, 20 July 1932, and J. S. Elias to Lord Beaverbrook, 21 July 1932, both in Beaverbrook Papers, Parliamentary Archives, London, BBK/C/299.

[59] "News in Brief," *Times* (London), 7 Apr. 1942, 2.

[60] W. A. Foyle, letter to the editor, *Times* (London), 6 May 1942, 5.

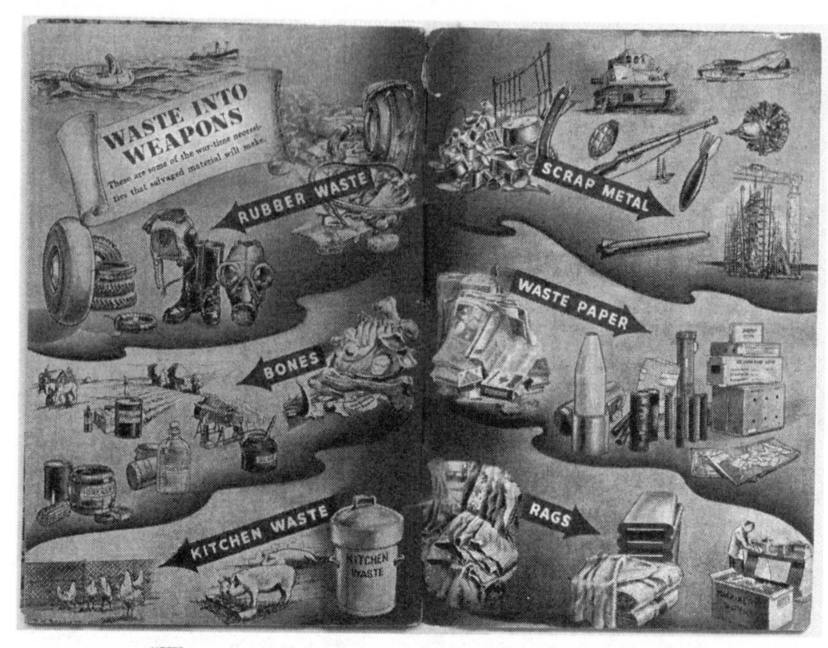

FIGURE 19. "Waste into Weapons." Using every form of media that existed, the Ministry of Supply and the Ministry of Information worked together to persuade the people of Britain that recycling was essential to victory. The metamorphosis of domestic articles into war materiel occupied a prominent place in this propaganda.
Source: The Amazing Story of Weapons Made from Waste (London: Odhams, 1942).

Foyle was fighting an uphill battle. The day after his warning appeared, Lady Mountbatten opened an exhibition at Selfridges, organized by WVS, which was designed to show people how waste paper was being converted into munitions. "By means of maps and statistics," explained the *Times*, "it is shown how during 1941 waste paper and cardboard were recovered which would have filled 150 ships. A striking mural painting by James Sharp describes the collection of paper . . . through to the final needs of the services, and has itself been painted on board made from waste paper. Examples of domestic salvage are shown with specimens of the equivalent munitions, such as the *Radio Times* and two cartridge cases which can be made from it." The exhibition also featured a paper shredder, and organizers invited the public "to bring their confidential documents, love letters, photographs, or anything they

FIGURE 20. "How Waste Materials Can Help in the Building of a Bomber." Throughout much of the war, British propagandists used images and texts in an effort to make people think of waste materials as valuable resources that could be made into weapons.
Source: The Amazing Story of Weapons Made from Waste (London: Odhams, 1942).

would like to contribute to the war effort but would prefer not to have seen by eyes other than their own."[61]

Responding to criticism about the destruction of worthy books, the Ministry of Supply worked with the Library Association to establish procedures for screening donations for valuable books.[62] When officials launched a new drive for books in October 1942 they emphasized that worthwhile volumes would be preserved from pulping.[63] Despite such promises, however, as late as 1943 many communities lacked scrutiny committees that could rescue valuable or historically significant publications from the mountain of printed material being sent for pulping.[64] And even when careful scrutiny occurred, very few donated books escaped the pulp mill. In Oxford, for example, of 417,905 books that residents contributed in early 1943, "scrutineers" saved only 497 volumes, mostly books printed in the seventeenth and eighteenth centuries, for transfer to the Bodleian Library.[65]

In the summer of 1943, the Ministry of Supply launched yet another campaign for books. The *Times*, which of course had a vested interest in increasing the supply of newsprint, heaped praise on the campaign and expressed admiration for "the cunning choice of a name for the effort. It is part of a general salvage campaign; and salvage has come to mean scrapping, dissolution, adaptation that involves destruction. But this is the National Book Recovery Campaign. It comes not to destroy but to recover."[66] The newspaper explained that "book lovers have been reassured by the activities of the Scrutiny Committee, composed (in each centre) of the local librarian, booksellers, antiquarians, and others with bibliophile knowledge. The presence of such scrutineers, through whose hands every book given must pass, is a guarantee that no volume which is worthy of preservation will be sent to the pulping mills."[67]

61 "From Waste Paper to Munitions: Exhibition to Point a Moral," *Times* (London), 8 May 1942, 7.
62 Library Association, Emergency Committee Minutes, 17 July 1942, Special Collections, UCL Library Services, University College London, Library Association Inventory 1 (Main Archive) Series D, Box 29.
63 "Book Salvage Campaign," *Times* (London), 26 Oct. 1942, 2.
64 "Salvage News," WVS *Bulletin*, Feb. 1943, 2.
65 "Salved Books for the Bodleian: Result of Oxford Collection," *Times* (London), 19 Apr. 1943, 2.
66 "Book Recovery," *Times* (London), 24 July 1943, 5.
67 "Books for Salvage and for Use: 31 Million Collected in Campaign," *Times* (London), 20 July 1943, 6.

Not everyone believed these reassuring words. In a 1943 letter to the *Times*, the anthropologist and psychical researcher Eric Dingwall complained that the promise of scrutiny was an empty panacea, because "what is 'valuable' to the research student may be 'valueless' to the bookseller or erudite bibliophile."[68] Charles Peat, parliamentary secretary to the Ministry of Supply, offered little hope for the types of publications that Dingwall had championed, such as Victorian periodicals and ephemera. He defined the three categories that scrutiny committees were using to sort books: "those suitable for services' reading; those suitable for restocking war-damaged libraries; and those which, having no entertainment or instructional value, were suitable for repulping to make munitions."[69]

BOOKS AND RECORDS AS INTELLIGENCE SOURCES

Records from businesses provided a wealth of military intelligence during the Second World War. Insurance companies that had sold policies before the war possessed detailed plans of German cities and businesses, which included information about construction materials and fire hazards. "These sources of information," noted one official, "are known to and used by [the] Air Ministry and War Office . . . and much use has been made of the town map files, and of insurance plans and reports on industrial risks for objectives purposes."[70] Amid the drives to scour the nation's bookshelves and attics for salvageable materials, some officials worried that valuable intelligence information might be destroyed. In December 1941, the officer in charge of the Admiralty's photo library wrote to the *Times*: "Sir,–Recent correspondence in your columns has stressed the importance of rescuing old photographs from the salvage collectors on the grounds of their historical interest. May I draw the attention of your readers to another aspect of this question?" He went on to note that the Admiralty urgently desired to borrow "photographs of foreign countries, showing harbours, ports, piers, jetties and other objects. . . . Our library also needs photographs of inland areas showing

[68] Eric J. Dingwall, letter to the editor, *Times* (London), 2 Aug. 1943, 5. I am grateful to my student Hannah K. Hicks for bringing Dingwall's work on spiritualism to my attention.

[69] "Fifty Million Books Collected: 5,000 Scrutineers at Work," *Times* (London), 20 Sept. 1943, 2.

[70] A. J. C. Luckly, British Insurance Offices as a Source of Economic Intelligence, n.d., NA, PRO 30/95/2.

the nature of terrain, roads, railways, bridges, power stations, oil installations, canals and other interesting subjects."[71]

The Ministry of Economic Warfare, the agency responsible for estimating Axis production capabilities and identifying targets for sabotage or aerial attack, initiated a similar campaign to recover information of military value from books that were donated to recover drives. In February 1942, the *Times* assisted this effort by publishing the following announcement:

> The Ministry of Economic Warfare needs foreign business and financial directories, statistical publications, official handbooks, guide-books, maps, plans, trade and exhibition catalogues, telegraphic address books, and street and telephone directories. Even old editions help to trace the former activities of concerns or individuals. All such publications relating to enemy, enemy-occupied or neutral countries issued within the last 10 years will be gratefully received at the Ministry . . . where all those of use will be selected and the remainder forwarded for salvage.[72]

Supporting the new campaign in an editorial, the *Times* predicted that it was likely to appeal to many members of the public who had hitherto been reluctant to hand over their books for recycling into munitions. "Now comes a new use and a new dignity for treasures which, far from being waste, are found to have, in their unpulped, individual condition, a hard, practical, business value in the waging of war."[73]

In September 1942, an official in the Research and Records Department of the Ministry of Economic Warfare circulated a list of foreign directories that it hoped to obtain. Among the titles included was *Das Adressbuch von Polen für Handel, Industrie, Gewerbe und Landwirtschaft* [The Directory of Business, Industry, Trade, and Agriculture in Poland].[74] No copies came to light in Britain, so the ministry made inquiries with the New York Public Library and the Library of Congress. When these efforts failed as well, an official in London contacted the British embassy in Washington with the thought that perhaps someone in

[71] W. R. Slessor, letter to the editor, *Times* (London), 12 Dec. 1941, 5.

[72] "Foreign Handbooks Needed," *Times* (London), 7 Feb. 1942, 2. On the uses of this information, see Uta Hohn, "The Bomber's Baedeker: Target Book for Strategic Bombing in the Economic Warfare against German Towns, 1943–45," *GeoJournal* 34, no. 2 (1994): 213–30; Michael Weaver, "International Cooperation and Bureaucratic In-Fighting: American and British Economic Intelligence Sharing and the Strategic Bombing of Germany, 1939–41," *Intelligence & National Security* 23, no. 2 (2008): 153–75.

[73] "Surrender Your Souvenirs," *Times* (London), 9 Feb. 1942, 5.

[74] W. O. Hassall, Short List of Principal Directories in Order of Importance Which We Need and Cannot Locate, 12 Sept. 1942, NA, PRO 30/95/6.

America's Polish community owned a copy.[75] At the same time that the British government was looking in the United States for printed sources on enemy targets, the U.S. government was seeking the assistance of British officials. In March 1943, a member of the Office of Strategic Services (forerunner of the Central Intelligence Agency) sought help in locating a set of Baedeker guidebooks on Germany and the countries that it occupied.[76]

That year the Ministry of Economic Warfare issued a guidebook of its own that included every German town with at least 15,000 residents – and some with even fewer. Intended not for tourists but for the Royal Air Force, it carried the ironic title *The Bomber's Baedeker*. Compiled from an extensive body of sources, including directories and business records gleaned from waste paper drives, it provided detailed information about German industrial sites, transportation systems, utilities, and food supplies.[77] Intelligence derived from business records helped experts in Bomber Command estimate the relative flammability of various districts and neighborhoods. Their calculations, which influenced the types and quantities of bombs that the British and the Americans dropped in Germany, contributed to the devastation of scores of cities and the deaths of 400,000 civilians.[78] In summer 1943, as British bombers were incinerating one German city after another, the minister of economic warfare commended the Library Association for helping to locate military intelligence among the books that it had recovered from salvage drives.[79]

LOSS

Although librarians and scholars bemoaned the loss of valuable books during the war, they were comforted by the knowledge that at least one copy of most works was likely to survive somewhere. No such hope existed for unpublished documents. In contrast to the system of scrutiny that it recommended for books and magazines, the Ministry of Supply

[75] P. L. Bushe-Fox to A. E. Ritchie, 20 July 1943, NA, PRO 30/95/5.

[76] Allan Evans to W. O. Hassall, 26 Mar. 1943, NA, PRO 30/95/6.

[77] Ministry of Economic Warfare, *The Bomber's Baedeker (Guide to the Economic Importance of German Towns and Cities), Part III: Survey of Economic Keypoints in German Towns and Cities (Population 15,000 and Over)* ([London]: Ministry of Economic Warfare, 1943), NA, FO 837/1315.

[78] Hohn, "Bomber's Baedeker," 229.

[79] Roundell Palmer, third Earl of Selborne, quoted in "Book Recovery," *Library Association Record* 45 (Oct. 1943): 174.

resisted similar safeguards for manuscripts. The information that it sent to local authorities in August 1943 in preparation for its national book recovery and salvage drive still made no mention of what to do with unpublished material that passed through the hands of salvage collectors. When a subsequent circular mentioned the possibility of recovering substantial quantities of waste paper from attics as a result of fire safety inspections, it likewise failed to mention that some of this material would possess historical significance.[80]

In 1944, Joan Wake urged the Ministry of Health to ask urban and rural district councils to exercise greater discretion in deciding which records to send for salvage. She also suggested that the Ministry of Health allow volunteers to examine nonconfidential papers collected in salvage drives so that valuable documents might be saved. The official who considered Wake's suggestions dismissed them with the curt comment that "no action is indicated at the moment."[81]

When officials finally agreed that certain classes of unpublished papers ought to be preserved, the instructions that they issued condemned to the pulp mills countless documents that historians would later deem to be of great value, such as local government records and the personal correspondence of ordinary men and women. Even when people attempted to preserve papers that they deemed valuable, the sheer mass of materials donated for salvage made it inevitable that many important papers would fall through the cracks and be sent on to the mills.

Despite the absence of a consistent screening process, preservationists managed to rescue a number of significant documents, including the 1541 charter by which Henry VIII had founded Peterborough Cathedral. In June 1942, Joan Wake organized an exhibition in Northampton of some of the most significant items that her group had rescued from the salvage piles.[82] In the course of a paper drive in the parish of Bayston Hill, near Shrewsbury, someone donated a volume of bound parchment estimated to be over seven hundred years old, which contained "handwriting, beautifully executed . . . presumed to be in Norman French." Fortunately, a vigilant volunteer spotted it and retrieved it from a stack of books sent to the paper mills.[83]

80 Ministry of Supply, Salvage Circular 105 (26 Aug. 1943), Ulverston Urban District Council Records, Cumbria Records Office, Barrow, BSUD/U/S/Box 10.

81 A. N. C. Shelley to F. Slator, 16 Mar. 1944, NA, HLG 102/91.

82 "Historical Records in Salvage: Exhibition of Treasures Recovered," *Times* (London), 3 June 1942, 7.

83 "'Find' in Waste Paper," *WTW*, 7 Dec. 1940, 9.

Wartime paper salvage contributed a great deal to the nation in material terms, but in terms of Britain's history and culture, the cost was high. The recycling of manuscripts and rare books during the war caused irreparable harm to the cultural and literary inheritance of the United Kingdom. We will never know the identity of many of the historical documents "salvaged" during the war. Most of what was recycled disappeared without any notice being taken in the press or in documents that have been preserved; only a few traces remain. Among the records of a firm of solicitors based in the southern English town of Battle is a small handwritten note indicating that the coronership records for the years 1868 to 1926 were handed over to be recycled on 6 March 1942.[84] A search of the Discovery catalog, available online through the National Archives, reveals dozens of references to documents that were "sent for salvage" during the Second World War. Although this list represents a tiny fraction of the tons of papers that disappeared, it provides a useful indication of the breadth of destruction that took place in the name of salvage: a large number of papers from Lord Halifax's estate in Yorkshire, eight volumes of admission records from the City of London Maternity Hospital dating back to 1769, records of the poor law guardians in Chichester, records from the asylums committee of the London County Council, records from the Suffolk quarter sessions, fourteen volumes of nineteenth-century court registers from Bolton, papers from the Maudsley Hospital, and thirty-two volumes of correspondence belonging to the Stepney board of guardians.[85]

[84] Note dated 6 Mar. 1942, Archives of Charles Sheppard and Sons of Battle, Solicitors, East Sussex Record Office, Brighton, SHE 2/8/10. I am grateful to Christopher Whittick for bringing this document to my attention.

[85] National Archives of England, Wales, and the United Kingdom, Discovery catalog, discovery.nationalarchives.gov.uk, accessed 19 Mar. 2015.

9

Requisition

> To win the war, we must devote more resources to it than the enemy, and since he will devote as many as he can, we too must put every man and every machine that can possibly be spared to the purposes of war.
>
> –Geoffrey Crowther [1]

Emphasizing the military utility of recycling, a writer for the magazine *Waste Trade World and the Iron and Steel Scrap Review* stated in February 1941 that the railings collected since the war began had provided enough metal to make "240,000 bombs, each 250 lb. in weight."[2] In contrast to announcements such as this, which trumpeted the valuable contributions that railings were making to the production of munitions, the archival record indicates that until mid-1941 officials considered railings to be important primarily for symbolic reasons. The scrapping of railings, in their view, provided dramatic demonstration that the government was taking forceful measures to defend Britain from invasion. As an official with the Ministry of Supply put it in an April 1942 letter to a counterpart in the Treasury, discussions about railings during the first two years of war had "had a political flavour and the supply case was not such that their collection was essential. . . . Since then, however, conditions have changed. As you are aware, the supply of scrap from the U.S.A. is severely restricted and demands on this side are increasing. It is essential in order to meet our iron and steel programme that all the scrap which can

[1] Geoffrey Crowther, *Ways and Means of War* (Oxford: Clarendon Press, 1940), 7.
[2] "Arms and the Scrap," *WTW*, 22 Feb. 1941, 2.

be recovered in this country should be obtained." Underscoring this point, he added that "cost . . . is no longer the deciding factor."[3]

When he took the helm of the Ministry of Supply in June 1941, Lord Beaverbrook immediately set out to increase the flow of iron ore and ferrous scrap to Britain's steel mills from both home and abroad. Railings were an obvious domestic source of iron, but as we have seen, a host of legal provisions made it difficult for people to surrender their railings voluntarily and even harder for the government to seize them.[4] Instead of being dissuaded by such complications, Beaverbrook forged ahead.[5] In August 1941, his ministry announced that "authority has been obtained by means of a Privy Council Order under the Defence Regulations to enable the Ministry . . . to take down and remove railings in possession of tenants and to protect tenants from any obligation to the landlords arising of the removal of such railings."[6] The legal basis for this action came from the Emergency Powers Defence Act of 1940. Commenting on this act in a report that he co-authored that year, a young American intelligence analyst named William J. Casey, who would serve decades later as Ronald Reagan's campaign manager and then CIA director, noted, "Traditional liberties of free-born Englishmen were cast aside on May 24, 1940, when in 150 minutes the Houses of Parliament passed, and the English King signed, a sweeping bill giving to the Cabinet plenary powers beyond any ever exercised in any free democracy." This law, "by subordinating every individual, rich or poor, every piece of wealth, every available plant and all manpower, raises the Prime Minister to a position of power as high as that of the German Chancellor."[7]

At the same time that Beaverbrook was preparing the groundwork for the requisition of railings, he expected that imports of iron and steel scrap would not only continue, but actually grow to meet Britain's expanding steel production. On 6 August 1941, the Iron and Steel Control asked the Ministry of War Transport to make ships available to bring an additional 30,000 tons of scrap metal each month from the United States.[8] Less than

[3] F. T. May to F. G. Lee, 4 Apr. 1942, NA, T 161/1405.

[4] J. C. Carr to R. Stanton, 26 Apr. 1941, NA, HLG 102/92.

[5] Lord Beaverbrook, Proposed Addition of Paragraph 3A to Defence Regulation 50, 25 July 1940, NA, CAB 75/12.

[6] "Tenants' Railings to Be Removed," *Times* (London), 9 Aug. 1941, 2.

[7] Leo M. Cherne, William J. Casey, and James L. Wick, "An Analysis of the Mobilization of Economic Resources for War in Germany and England," Research Institute of America, [1940], 55, Stettinius Papers, Box 87.

[8] W. Graham to C. R. Morris, 7 Aug. 1941, NA, MT 59/865.

two weeks later, however, these plans changed abruptly. U.S. demand for iron and steel scrap had grown rapidly over the preceding twelve months, and its industry was "operating on a day-to-day basis," with few if any reserve supplies to serve as a buffer against disruptions in transportation or to allow for America's rapidly expanding military production.[9] Consequently, when Beaverbrook met with Roosevelt and Hopkins at the White House on 19 August 1941, the Americans informed him that they had decided to suspend exports of both scrap metal and pig iron to Britain.[10]

The very next day F. T. May of the Ministry of Supply wrote to a colleague in the Treasury. Lord Beaverbrook, he began, "has directed that a complete scheme should be prepared for the requisition of railings throughout the country, and it is probable that this will be launched within the next two weeks." May did not explain the reason for the new policy, but the timing was more than coincidental. As another official put it a few weeks later, Beaverbrook's decision to requisition privately owned railings was a direct "consequence of the discontinuance of the supply of scrap from America."[11]

If owners requested compensation for their railings, May suggested that the government should offer at least 40 shillings (£2) per ton, as anything lower than that would seem insulting. He admitted, however, that such compensation, when added to the expense of removal and transport, would exceed the price that the government had established for scrap. To minimize a massive drain to the Treasury, May hoped that most owners would give up their railings without compensation. "Forms of claim," he added, "will only be issued by local authorities on application."[12]

Responding to this letter, the Treasury official F. G. Lee asserted that people "who have not been patriotic enough to offer their railings to the nation" did not deserve the "generous" compensation that May had

[9] "Lord Beaverbrook Sees the President," *Times* (London), 20 Aug. 1941, 4; *Institute of Scrap Iron and Steel Yearbook, 1942*, 20–1.

[10] "Scrap and Pig-Iron – 1942," 24 Dec. 1941, Beaverbrook Papers, Parliamentary Archives, London, BBK/D/111. This version of the document consists of six numbered paragraphs and is signed by C. R. Morris; an adjacent version contains three short paragraphs and is unsigned.

[11] F. T. May to F. G. Lee, 20 Aug. 1941, NA, T 161/1405. On the linkage between the U.S. suspension of scrap exports and the British decision to requisition privately owned railings, see E. J. R. Edwards to Treasury, 17 Sept. 1941, NA, T 161/1405.

[12] F. T. May to F. G. Lee, 20 Aug. 1941, NA, T 161/1405.

proposed. "If a few shillings (or even some smaller sum) is all that the claimants are legally entitled to . . . that is all that they should receive; and we feel that it is definitely wrong for the Ministry of Supply to make an offer on a basis which they know will involve public funds in loss." Instead of offering 40 shillings per ton, Lee suggested that 25 or 30 shillings would be a more appropriate amount. The difference between this amount and the price the government received from industry for the severed railings, he explained, would pay for their removal.[13] Although imported steel scrap cost around £10 per ton during the war, the government decided to sell railings to the scrap trade for only 62s. 6d. per ton (slightly more than £3). A higher rate would have generated more revenue, but it also would have increased the cost of everything made from steel.[14]

If an owner refused to accept compensation at the standard rate of 25 shillings a ton, he or she would have the right to appeal to a tribunal. The Compensation (Defence) Act, 1939, which governed all types of requisitions during the war, allowed property owners to make claims on two grounds. Section 3 provided restitution for a decline in the value of land, but this was not to include "any such diminution ascribable only to loss of pleasure and amenity" – a phrase that would void any claims in respect to railings that served a purely aesthetic function. Section 6 allowed owners to claim reimbursement for the value of removable articles taken by the government. In the case of railings, officials cynically decided that the value would be assessed immediately after, rather before, they were detached from the ground.[15] "As I understand it," commented one Treasury official in August 1941, "the Ministry proposes to pull down railings and take the line that scrap iron value only is payable for them. I very much doubt whether the Tribunal will share that view. I think the least compensation they would be likely to give is the value of the railings as railings not as scrap iron. How far the Ministry are likely to get away with the payment of scrap iron value is I think a matter for you to judge." In the margin beside these words one civil servant added, "This is worrying; no use, I suppose, to try another public appeal?"[16]

[13] F. G. Lee to F. T. May, 26 Aug. 1941, NA, T 161/1405.

[14] A year before the government began to seize them, one official had warned that the expense of removing donated railings would often exceed their scrap value. See C. J. Pyke to F. G. Lee, 6 Aug. 1940, NA T 161/1405.

[15] Report: Requisitioning of Railings, 21 Aug. 1940, NA, HLG 102/92.

[16] H. M. Young to J. S. Bromley [handwritten annotation changed this to F. G. Lee], 25 Aug. 1941, NA, T 161/1405.

As the government prepared to take people's railings, one Treasury official quipped that the new policy provided "a real case for propaganda experts to show their paces. It is my belief that the majority of owners, so far from wishing to haggle over a possession which has long ceased to be of real importance (and the proposal expressly excludes railings which are still necessary or pleasing) will be stimulated by this evidence that the Government is leaving no stone unturned to win the war of production."[17] The argument that owners would be happy to have the government liberate them of their railings was extremely disingenuous, because officials never intended to give owners the choice of whether they wanted to keep them. Furthermore, as had already become clear in the case of railings that owners had relinquished voluntarily, removal often caused a great deal of damage to the stones in which they were set.

On 11 September 1941, the Ministry of Supply sent a two-page letter to every local government body in Great Britain (but not Northern Ireland, which was treated differently when it came to salvage). The letter began by stating that the minister of supply had "decided by the exercise of his powers under the Defence Regulations to requisition all unnecessary iron and steel railings in the country for scrap for use in iron and steel works and foundries." The memorandum directed government officials in each city and town to survey all public and private railings, gates, posts, chains, and bollards within their jurisdictions and to submit a report within six weeks, detailing their locations and approximate weights. The only such objects to be retained were those that were essential for safety, possessed special artistic or historical significance, or were needed to keep livestock from straying.[18] Two weeks later, the ministry declared that railings and gates should not be taken if some other type of material would be needed to replace them.[19] Thus began one of the most remarkable instances of the power of the British state in wartime. Over the next three years, government representatives removed and melted down nearly all of the nation's railings.[20] During the first year of requisition, officials took 310,315 tons of iron railings from 2 million houses in Britain.[21]

[17] J. S. Bromley to F. G. Lee, 23 Aug. 1941, NA, T 161/1405.
[18] R. D. Fennelly, circular letter to local authorities, 11 Sept. 1941, NA, T 161/1405.
[19] R. D. Fennelly, circular letter to local authorities, 25 Sept. 1941, Ulverston Urban District Council Records, Cumbria Records Office, Barrow, BSUD/U/S/Box 10.
[20] H. W. Secker to A. W. S. Edwards, 6 Apr. 1944, NA, WORK 22/430.
[21] *Parliamentary Debates*, Commons, 30 Sept. 1942, vol. 383, cols. 764–5.

In marked contrast to many of the other materials that the government required people to recycle in wartime, railings were clearly not rubbish. Consequently, many owners expressed bewilderment and outrage when officials took them as a source of scrap metal. During and after the war, critics inside and outside the government charged that the railings requisition program was unjust, destructive, capricious, and lacking in transparency. To add insult to injury, the Control of Timber (No. 24) Order made it illegal for owners to replace requisitioned iron gates or railings with wooden structures.[22]

Shortly after the Ministry of Supply announced its requisition program, the author of an article in a prominent architectural journal called for the preservation of railings that were integral to a building's overall design. Despite his spirited defense of quality railings, the writer had nothing but contempt for ordinary ones, claiming, "We would gladly part with the great quantity of cast-iron gates and railings, crude in design and execution, to be found enclosing the front gardens of small houses in suburban areas of the late nineteenth and early twentieth centuries. Let all this be removed before we begin to destroy ironwork of merit, whether ancient or modern."[23] In both its reasoning and its language, this argument displayed an uncanny resemblance to Neville Chamberlain's willingness at Munich to sacrifice territory that belonged to the recently established and "far-away country" of Czechoslovakia.

Following the deeply ingrained British pattern of seeking formal consent (even if obtained through coercive methods, such as in the campaign to persuade men to enlist during the first two years of the Great War) the Ministry of Supply tried to create the pretense that the railings it took were surrendered voluntarily. This approach served the financial interests of the Treasury, for if owners could be made to think of their loss of railings as a gift to the nation, they would be unlikely to ask for compensation.[24] Other officials pointed out that because only "unnecessary" railings were taken, it would be illogical to claim that their removal entailed a monetary loss or that any need existed to replace them.[25]

[22] Ministry of Works and Planning, memorandum, 15 Oct. 1942 (copy), NA, AO 30/32.

[23] "Railings for War Weapons," *Builder*, 17 Oct. 1941, 344.

[24] Raw Materials Department to T. S. Overy, 12 Sept. 1941 (copy), NA, POWE 5/59. On the coercion of potential recruits between 1914 and 1916, see Nicoletta Gullace, *"The Blood of Our Sons": Men, Women and the Renegotiation of British Citizenship during the Great War* (New York: Palgrave Macmillan, 2004).

[25] Ministry of Works, Directorate of Lands and Accommodation, Emergency General Instruction No. 206, 14 Feb. 1942, NA, AO 30/32.

To generate public support for its decision to requisition the nation's gates and railings, officials orchestrated an elaborate public relations campaign. In early September 1941, an official with the Ministry of Works wrote to the king's private secretary with a request to remove a substantial quantity of metalwork from Buckingham Palace, consisting of a pair of large gates from the Royal Mews, various internal railings, and "the portions of the main front railings which were damaged by enemy action." In his letter, the official explained that it was not the weight that mattered, but rather "the royal lead that would thus be given to the campaign for scrap."[26] The king quickly agreed, and the Ministry of Works trumpeted his generosity in a press release, which explained that these railings and gates, weighing some twenty tons, would "be converted into scrap for the manufacture of tanks and other war weapons."[27]

The ministry carefully choreographed the removal of railings from the palace, and many members of the press came to record the spectacle. The man who took the first crack at them was George Hicks, a former bricklayer and leader of the Trades Union Congress who had been the Labour MP for East Woolwich since 1930. A decade later he became parliamentary secretary to the Ministry of Works, a position that he would hold throughout the duration of the war in Europe. Hicks, who loved publicity, seemed to relish the symbolism inherent in breaking down part of the protection around the palace. Sporting welding glasses and heavy leather gloves, Hicks cut through the first section of railings as sparks flew all around. Dutifully covering the event, the *Times* reported that the Ministry of Works expected to haul away twenty tons of iron "before the end of the day. It will be kept together as a lot, and may be used to provide a 'Buckingham Palace' tank."[28]

Although many people thought that all railings should be melted down, others called for restraint. In September 1941 Sir Eric de Normann, Britain's top official in charge of parks and public buildings, told his colleagues at the Ministry of Works that he opposed

[26] Geoffrey Whiskard to Alexander Hardinge, 6 Sept. 1941 (copy), NA, WORK 19/1141. In contrast to much contemporary practice, this letter did not capitalize the word royal.

[27] Alexander Hardinge to Geoffrey Whiskard, 17 Sept. 1941, NA, WORK 19/1141; Ministry of Works, "Press Notice: Removal of Railings from Buckingham Palace," 1 Oct. 1941, NA, WORK 19/1141.

[28] "Removal of Palace Railings: The Work Started by Mr. George Hicks," *Times* (London), 4 Oct. 1941, 2; Eric de Normann, "Hicks, (Ernest) George (1879–1954)," rev. Marc Brodie, in *Oxford Dictionary of National Biography* (Oxford: Oxford University Press, 2004).

surrendering any more park railings where these serve a useful function or where there is any possibility of our having to restore them after the war. For the moment when one looks round one sees immense quantities of steel girders still available in damaged buildings and as these provide by far the easiest and most suitable supply of scrap they should be taken first. [L]ater on if we are reduced to extremities we shall have to consider scrapping more park railings, but for the moment we have already set a very handsome example by giving up some 1,200 tons of the more easily spared railings.[29]

A short time later another civil servant, E. J. R. Edwards, urged de Normann to take more drastic action. Edwards argued that the government had to set the tone for the sacrifices that it was imposing on the public by being ruthless in regard to its own railings:

The attitude of individuals and of local authorities whose railings we propose to take will obviously be influenced very considerably by the attitude of Government Departments in this matter: and the Bailiff of the Royal Parks, I suggest, is in the unique position of Caesar's wife. A glance at the papers in this file would seem to show that the sacrifices already made will still leave a large number of railings, the need for whose retention will not be obvious to the public, and so a bad impression will be created.... In view of the psychological value to the railings campaign of sacrificing the perimeter fencing of (say) St. James' [*sic*] Park, I suggest very seriously that you reconsider your dictum that we must be able to close the parks.

Responding to de Normann's arguments that blitz scrap should be taken first, Edwards observed that other considerations were more important: "It is a question of a striking gesture which would show that the Ministry of Works, as the main Governmental property-holder, is going to be as generous as it expects the ordinary citizen to be."[30]

Rejecting Edwards's insinuation that he was failing to pull his weight, de Normann declared that "we are probably the only undertaking which has carried out a complete schedule of our resources and which has already given up a very large percentage of them. But I maintain that the release of railings must really follow some coherent and economical plan." He added that the Ministry of Supply had promised not to remove railings that served a useful purpose. "This," noted de Normann, "certainly covers the Park railings which would have to be replaced after the war. They are wanted for all sorts of purposes: the preservation of public

[29] Eric de Normann to Ministry of Works, 16 Sept. 1941, NA, WORK 16/1710. Appointed to the Office of Works in 1920, Eric de Normann (1893–1982) had responsibility for both public buildings and parks. See "Sir Eric de Normann," *Times* (London), 28 Jan. 1982, 14.

[30] E. J. R. Edwards to Eric de Normann, 15 Oct. 1941, NA, WORK 16/1710.

decency, the protection of public property in the Park, e.g. many thousands of chairs, valuable plants and bulbs and for restraining crowds on ceremonial occasions, etc."[31]

At this point the fate of railings around the royal parks became the subject of a power struggle between Lord Beaverbrook, who had become minister of supply in June 1941, and Lord Reith, the minister of works. On 21 October 1941, Beaverbrook sent an impudent letter to Reith:

My dear John,

I require the railings from the Royal Parks. This would set a good example to private people.

They are scheduled by the local authority for removal. The Ministry of Supply has the right to give the direction for them to be removed.

Will the Ministry of Works and Buildings please carry out the direction forthwith. If you have any objection I will ask the proper department in this Ministry to take action.

Yours ever,
Max[32]

Just one day after Reith received this letter, Beaverbrook's secretary rang to ask how soon the minister would reply. This action prompted a sharp rebuke from a senior member of Reith's staff, who reminded the caller that neither local authorities nor the Ministry of Supply had the authority to requisition royal property.[33] Reith's reply to Beaverbrook was even more abrupt and did not even contain a salutation. In it, he noted that at the request of the general officer commanding the London district, the railings around Kensington Gardens, Hyde Park, St. James's Park, and Green Park would remain as a defense against invasion. He noted, however, that the railings that surrounded "Greenwich and Regent's Park[s] will go assuming The King agrees."[34]

George VI soon gave his consent to this proposal, but he hoped that no additional parks would have to lose their boundary railings. As his private secretary, Sir Alexander Hardinge, put it in a letter to the Ministry of Works, "The King has no objection to the removal of as many internal railings as possible in the different Parks, as well as the external railings at Regent's and Greenwich Parks. At the same time His Majesty cannot understand why railings which are serving some useful purpose are turned

31 Eric de Normann to E. J. R. Edwards, 17 Oct. 1941, NA, WORK 16/1710.
32 Lord Beaverbrook to Lord Reith, 21 Oct. 1941, NA, WORK 16/1710.
33 "Park railings," 27 Oct. 1941, NA, WORK 16/1710.
34 Lord Reith to Lord Beaverbrook, 28 Oct. 1941 (copy), NA, WORK 16/1710.

over to scrap, instead of [the Ministry of Works] collecting the mass of material still left in buildings that have been bombed."[35]

Two days later Beaverbrook again sought to take all the railings that remained in the royal parks. At a meeting of the Defence Committee on 5 November, he asked whether the military remained opposed to the removal of the railings that surrounded the parks of central London. They had favored keeping them as a way to impede the advance of German parachute troops in the event of invasion. After Churchill suggested that barbed wire could achieve the same ends, military officials dropped their opposition to the removal of park railings. The next morning, convinced that he had outmaneuvered Reith, Beaverbrook called the cabinet secretary, Sir Edward Bridges, and boasted that he had "got away with it" the previous evening. Beaverbrook said that he wanted to make sure that Bridges included the decision in the official minutes of the meeting. Bridges refused and told Beaverbrook that "it was not a decision but an expression of opinion by the people present, who did not include everyone concerned." The cabinet secretary reminded Beaverbrook that the king, as well as Reith, had to agree to the removal of any royal railings. Bridges tried to telephone Reith to let him know what was happening but, as Reith was unavailable, he spoke with one of Reith's subordinates. Bridges assured him that the minister of works "was still entitled to make his own decision and if necessary take the matter to the Cabinet."[36]

Evidently Reith gave in without contesting the issue, for the next day one of his senior officials wrote to Alexander Hardinge, private secretary to King George VI. The letter sought the king's permission to remove all of the railings that still surrounded the royal parks, which "the Ministry of Supply are pressing us to cut down . . . for scrap."[37] Four days later Hardinge replied, "The King agrees to the removal of all iron railings from all the Parks in view of the fact that both the Military and the Police Authorities are satisfied that it will have no evil effects."[38] By the end

[35] Alexander Hardinge to Geoffrey Whiskard, 3 Nov. 1941, NA, WORK 16/1710. Extracts from this letter are reproduced with the kind permission of Her Majesty Queen Elizabeth II. On Hardinge's role, see Hugo Vickers, "Hardinge, Alexander Henry Louis, Second Baron Hardinge of Penhurst (1894–1960)," in *Oxford Dictionary of National Biography* (Oxford: Oxford University Press, 2004).

[36] Note by F. J. Root, 6 Nov. 1941, NA, WORK 16/1710.

[37] Geoffrey Whiskard to Alexander Hardinge, 7 Nov. 1941 (copy), NA, WORK 16/1710.

[38] Alexander Hardinge to Geoffrey Whiskard, 11 Nov. 1941, NA, WORK 16/1710. Extracts from this letter are reproduced with the kind permission of Her Majesty Queen Elizabeth II.

of 1941, workmen had removed miles of railings from London's royal parks, including Hyde Park, Kensington Gardens, St. James's Park, and Green Park – and work on those of Regent's Park was slated to begin shortly. At the same time, the Corporation of London, which governed the "square mile," announced that it would soon take railings from St. Paul's Churchyard, Postman's Park, the Mansion House, and many other prominent locations in the City of London.

Aware of the symbolic significance that railings had assumed in wartime, on 28 October 1942 the Soviet ambassador, Ivan Maisky, used a welding torch to cut through the railings in front of his country's embassy in Kensington Palace Gardens. These railings, explained the *Times*, were being given freely by the USSR, as diplomatic privilege exempted them from the requisition order. Speaking to the reporters who had come to witness the spectacle, the ambassador declared, "Every bit of scrap metal should be put into the melting pot, and I am glad to help in this small way to increase the aid that is received by my country."[39] Maisky's enthusiasm soon soured, however, after members of the public began to trespass on the newly accessible grounds of the Soviet embassy. After looking into the matter, an official in the Ministry of Works recommended allocating funds to build a replacement wall. "Although we do not normally provide a substitute for railings, we feel that we cannot allow an incident to develop between ourselves and the Russian Government over this small matter." After discussing potential designs with the ambassador's wife, the ministry agreed to spend £250 to construct a wall where the railings had stood.[40]

The government's order to schedule "unnecessary" railings for removal caused considerable confusion and disagreement, in large part because it involved so many agencies at both the national and local levels. Although requisition policy originated in the Ministry of Supply, local volunteers or municipal officials compiled inventories, private contractors carried out the work of removal and transport, and the Ministry of Works paid the contractors. To complicate matters further, even though the government gave local officials the responsibility of determining which railings should be saved, the Ministry of Works had the final say.[41]

In their zeal to collect railings, officials quickly backed away from their promise to spare those that were artistically or historically significant. When L. L. Dussault, the architect commissioned to inspect railings in

[39] "Soviet Embassy Railings for Munitions," *Times* (London), 29 Oct. 1942, 2.
[40] R. M. Hunter to H. E. C. Gatliff, 30 Jan. 1943, NA, T 162/748.
[41] "Nation-Wide Search for Scrap Iron," *Times* (London), 21 Jan. 1942, 8.

FIGURE 21. Soviet ambassador cuts railings of embassy. Ivan Maisky, Soviet Ambassador to the United Kingdom, donated the iron railings that surrounded his country's embassy in Kensington Palace Gardens to assist the British drive for scrap iron. On 28 October 1942, with curious members of the press and public gathered around him, he began removing them. Maisky soon came to regret the loss of these railings, however.
Source: Photo by Popperfoto. Courtesy of Getty Images (79667779).

Stratford-upon-Avon, filed his report in December 1941, he recommended the removal of a pair of gates at the riverside entrance to the Shakespeare Memorial Theatre (now known as the Royal Shakespeare Theatre). Although he admitted that they were "artistic," he ruled that their "recent manufacture" meant that they had to go.[42] Three months later, Dussault likewise ruled that gates belonging to three local churches should be removed: "I . . . cannot pass them as antique; they may be called artistic, but we cannot retain such."[43]

[42] L. L. Dussault to P. C. Smart, 15 Dec. 1941, Shakespeare Centre Library and Archive, Stratford-upon-Avon, BRR 55/14/31/5/3.
[43] L. L. Dussault to P. C. Smart, 13 Mar. 1942, Shakespeare Centre Library and Archive, Stratford-upon-Avon, BRR 55/14/31/5/3.

Following complaints that local authorities were failing to carry out requisition consistently throughout Britain, the Ministry of Works issued detailed instructions about the types of railings that should be retained. These included railings at schools that were needed "to prevent young children running on to a highway or to prevent misuse of outside offices [lavatories]," those that protected military installations and munitions factories, those at "mental hospitals, lunatic asylums and hospitals for infectious diseases," and "railings round burial grounds and cemeteries in rural areas that are not set on an external wall of a minimum height of 3 feet from ground level." The last exemption was likely intended to prevent cattle from wandering into rural graveyards and knocking over tombstones. Although officials hoped that these instructions would make requisition more consistent and more accepted, the new document failed to include the earlier directive to preserve railings that possessed special artistic merit or historical character. In the words of the circular, "Many enquiries have arisen out of special pleas for individual railings and gates and it may assist Local Authorities to know that generally apart from the typical cases mentioned no such pleas should be considered."[44]

Many people assumed that the government would notify them before seizing their property, but this did not occur. Instead, officials placed notices in newspapers, which informed owners that they had fourteen days in which to appeal if they objected to the removal of their railings. Although most people no doubt heard of the government's plans to requisition unnecessary railings, many of them considered it self-evident that their railings served a useful purpose – and were therefore exempt from requisition. When this assumption proved incorrect, owners often experienced feelings of incomprehension, anger, and betrayal. In the face of the complaints that followed, officials responded that the burden rested on property owners to file a preemptive appeal, not on the government to notify citizens whether and when it would take their railings. "If owners failed to appeal after the public notice appeared," declared one local surveyor, "it was assumed that they had no objections."[45] In numerous cases, work crews took railings even after local officials had determined that they should remain. As one surveyor observed, the contractors

[44] Ministry of Works, circular letter to local authorities, 9 Jan. 1942, NA, HLG 102/94. Copies of this circular also exist in NA, ED 138/64 and HLG 102/92.

[45] For an interesting view of the frustration that this confusion caused, see A. N. C. Shelley, Railings for Scrap, 5 Jan. 1942, NA, HLG 102/92.

employed by the Ministry of Works went about their task with an "attitude . . . of ruthlessness. . . . They decided to take railings and argue afterwards. For that reason many railings are being taken down to-day that I did not schedule."[46]

As soon as railings were severed, they became the property of a government-owned holding company called the British Iron and Steel Corporation, which the Ministry of Supply had established on 11 June 1940. The corporation financed the collection not only of railings but also of "blitzed" materials, tramlines, obsolete buildings, and old equipment, and facilitated the sale of recovered materials to the scrap trade. Despite the paltry compensation it gave to those who requested it and the fact that most owners did not ask for any compensation, the corporation operated at a loss due to the high costs of removing and transporting scrap. By the end of end of 1943, the corporation had a deficit of £1.2 million, nearly half of which came from the requisition of railings.[47] Although this was a large sum, it was far less than the cost of importing a similar quantity of scrap.

By the time the United States entered the war in December 1941, fewer than half of the local authorities that had been ordered to compile inventories of private railings had done so.[48] Those that had failed to comply received a sharply worded letter from G. M. Carter, the director of demolition and recovery at the Ministry of Works. In January 1942, he wrote to the town clerk of Walsall to point out that its council had yet to submit a list of railings to be removed. "I am instructed by Lord Reith to ask that steps shall be taken forthwith to furnish the schedules required. . . . Lord Reith is most anxious that with the growing demand for scrap metal for the War effort in consequence of the extension of the War to the Pacific every effort shall be made to facilitate the work which he has undertaken."[49] Towns and cities that neglected their responsibilities would face stiff fines.

APPEALS AND COMPLAINTS

To determine which railings should be preserved, the Ministry of Works turned to the local panels of architects that had been set up to decide the

[46] "Railings Protest," *Public Cleansing*, Apr. 1942, 224.

[47] Ministry of Supply, Memorandum No. 315/44, 28 Oct. 1944, POWE 5/114.

[48] "National Scrap Survey: Much More Needed," *Times* (London), 15 Dec. 1941, 2.

[49] G. M. Carter to town clerk, 22 Jan. 1942, Walsall Local History Centre, 235/4.

fate of bomb-damaged buildings. Their mandate was to review "schedules of railings prepared by Local Authorities with a view to the deletion of railings of special artistic merit or historic interest." These panels would also consider appeals from the public. Their decisions, if unanimous, were final. When cases could not be resolved at the local level, the ministry referred them to its "Headquarters Panel" of four experts: C. T. P. Bailey (keeper of ironwork at the Victoria and Albert Museum), Professor A. E. Richardson (a champion of eighteenth-century design), and two members of the Inspectorate of Ancient Monuments, H. M. Fletcher and George Hulbert Chettle.[50] The Ministry of Works directed this panel to be very strict when considering appeals from owners who objected to the removal of their railings.[51]

In October 1941, the Ministry of Works informed local government officials that "the railings which it is proposed to retain . . . should only be outstanding examples of their type or period. It may be assumed for this purpose that only specimens of the finest craftsmanship and design should be retained and that generally speaking no railings manufactured since 1820 come within this category. In other words, railings that can reasonably be expected to be reproduced after the war must be sacrificed to the urgent national necessity for scrap."[52] The effect of this slight change in interpretation was enormous: no railings, regardless of how expensive or finely crafted, were safe from the furnaces. Unfortunately, the ministry failed to announce this policy change to the public.

Those who sought to preserve their railings faced not only legal hurdles but also public opprobrium. When Hastings Russell, twelfth Duke of Bedford, a major landowner in London, objected to the planned requisition of the railings that surrounded Russell Square, many people denounced him as unpatriotic. Others went even further. In October 1943, cloaked by the nightly blackout, vandals defaced the statue of the duke's grandfather that stood in the square. Using yellow paint, they wrote, "Down with the Duke and the Railings" and added the words

[50] Ministry of Works and Buildings, circular letter to local authorities, 22 Oct. 1941, Walsall Local History Centre, 235/4; R. M. Hunter to H. E. C. Gatliff, 8 Nov. 1941, NA, T 162/690; "Preservation of Iron Railings of Interest," *Builder*, 14 Nov. 1941, 434; "Preservation of Iron Railings of Architectural and Historic Interest," *Journal of the Royal Institute of British Architects* 49 (Nov. 1941): 16; "Mr. G. H. Chettle: Protecting Ancient Monuments," *Times* (London), 27 Sept. 1960, 9.

[51] "Removal of Railings," *Municipal Review*, Feb. 1942, 23.

[52] Ministry of Works and Buildings, circular letter to local authorities, 22 Oct. 1941, Walsall Local History Centre, 235/4.

"melt me" beside the statue. On the other side of Russell Square someone used chalk to write, "Give us these railings" on the walking path.[53] The protestors soon got their wish, for in early November the Ministry of Works removed the railings from Russell Square.[54]

Another prominent individual who came under fire for trying to save his ironwork was the former prime minister, Stanley Baldwin. Keenly aware of the hostility that the Duke of Bedford's actions had aroused, Beaverbrook intervened directly to ensure the removal of ornamental railings from Baldwin's estate, Astley Hall. Even this was not enough to satisfy some critics, for after the Ministry of Works decided to leave standing a pair of custom-made gates that Baldwin had received as a gift from his supporters upon his retirement from politics, one member of parliament sarcastically suggested that they were necessary "to protect him from the just indignation of the mob."[55]

The *Times* reported in February 1942 that officials had recently reversed their earlier decision to melt down the railings that stood outside St. Paul's Cathedral, designed in the early eighteenth century by the Huguenot immigrant Jean Tijou, as well as those from Samuel Johnson's house in Gough Square and others in Cheyne Walk, a picturesque street in Chelsea famous for the large number of writers, artists, musicians, and politicians who have lived there.[56] Although this news no doubt pleased preservationists, the fact that these railings had nearly been scrapped no doubt caused alarm. Less fortunate were the railings from St. Martin-in-the-Fields, St. Clement Danes, and seven other London churches, which the Ministry of Works boasted would yield "enough scrap iron to make at least six medium-sized tanks."[57]

Decrying such destruction, the novelist Evelyn Waugh went so far as to describe railings as among the "unique achievements" of the Victorian era. Amid "the high mood of sacrifice with which so much is being broken and melted," he detected "an undercurrent of satisfaction that national need should give the opportunity for removing what is now thought unsightly." Waugh further argued that the widespread desire to see railings disappear

53 "Down with the Railings," *Times* (London), 1 Nov. 1941, 6; Richard Griffiths, "Russell, Hastings William Sackville, Twelfth Duke of Bedford (1888–1953)," in *Oxford Dictionary of National Biography* (Oxford: Oxford University Press, 2004).

54 "News in Brief," *Times* (London), 11 Nov. 1941, 2.

55 "Lord Baldwin's Gates," *Times* (London), 6 Feb. 1942, 2; "House of Commons," *Times* (London), 5 Mar. 1942, 8.

56 "Care in Scrapping Railings: London Exemptions," *Times* (London), 3 Feb. 1942, 2.

57 "Brevities," *Devon and Exeter Gazette*, 31 Oct. 1941, 5.

represented hostility to the ideas that they embodied. "The railings which adorned the homes of all classes were symbols," he suggested, "of independence and privacy . . . in an age which rated liberty above equality. The prevailing sentiment deprives them of this value." He observed that each generation tended to denigrate the fashions of its predecessors, and he reminded his readers that "it is only recently, still imperfectly and after heavy losses, that the work of the eighteenth century has been recognized as having aesthetic value." In an analogous manner, he predicted that someday "we and our children will look back with increasing curiosity to the free and fecund life of Victorian England; may we therefore ask that the responsible officials will consider the inclusion of a few relics of the period among those they are protecting from the scrap heap?"[58]

Affirming much of what Waugh had written, the editors of the *Times* asserted that "it is a sound principle that no age has the right to destroy the art of any previous age on aesthetic grounds alone." Contrary to what most people believed, they suggested that "there must be history, there may even be poetry . . . in what our own time feels to be the very ugliest of Victorian domestic stuff. Should the rustic seats and the cast-iron fountains never (as Captain Waugh supposes they may) come back into favour, they have a right to their place in the history of our civilized life." Despite this support, the editorial ended by noting that the brutal logic of war might require their destruction. Employing a flurry of double negatives, it asserted that "we cannot pretend that national existence without Victorian railings is not better than Victorian railings without national existence."[59]

Across the United Kingdom, thousands of people protested the government's plans to take their property. The Shakespeare Centre in Stratford-upon-Avon holds a particularly rich collection of such letters. Referring to the safety exemption, R. Summerton informed the town council that his railings were necessary to prevent passersby from falling into the cellar light of his house, and he vowed to hold the town "liable for any accidents caused by the removal of the said railings."[60] Four days later, the borough engineer and surveyor of Stratford replied that officials would use sandbags to ensure that no such hazard existed. He neglected to mention how

[58] Evelyn Waugh, letter to the editor, *Times* (London), 3 Mar. 1942, 5.
[59] "Taste in Transition," *Times* (London), 3 Mar. 1942, 5.
[60] R. Summerton to town clerk, 12 June 1942, Shakespeare Centre Library and Archive, Stratford-upon-Avon, BRR 55/14/31/5/2.

this would affect the amount of light that could reach Summerton's basement windows.[61]

Some owners argued that their gates or railings contained so little metal that the government would spend far more in removing them than they were worth as scrap. To promote a sense of shared sacrifice and equal treatment, however, officials rejected such pleas and seized everything they could, even though, as one of them confided at a meeting with Treasury officials, "there would be many which were not worth the labour of collecting. It was nevertheless the Ministry's intention," explained R. M. Hunter, "to avoid selection in order to avoid the criticism and argument which would be caused by such selection." These words flatly contradicted the government's repeated claims that it would requisition only unnecessary railings that had no value other than as scrap. Furthermore, Hunter's later remarks at the meeting revealed that officials treated wealthy property owners more favorably than ordinary people. To avoid "very high compensation claims" from the owners of valuable railings, he explained that members of his staff were "revising local authorities' schedules here and there."[62]

The requisition of railings applied not only to houses but also to a wide range of other properties. When civil servants in the Board of Education learned that the Ministry of Supply planned to requisition railings from schools, they expressed frustration that they had not been consulted. One of the things that most angered them was the ministry's announcement that contrary to its previous assertions, appeals against the requisition of railings would be considered only on the basis of architectural merit or historical significance, not on the basis of safety.[63]

Those who appealed against the removal of railings from schools and playgrounds often argued that their removal would endanger children. In December 1941, for example, the secretary of the Guinness Trust complained that "much concern is being felt among owners of industrial dwellings [i.e., low-income housing] at the wholesale removal of railings surrounding their properties. In a particular case in point, all the railings were removed recently from three frontages of a two-acre site belonging to this Trust; all three frontages are on busy thoroughfares, and there is

[61] P. C. Smart to R. Summerton, 16 June 1942, Shakespeare Centre Library and Archive, Stratford-upon-Avon, BRR 55/14/31/5/2.

[62] "Note of Meeting Held at Treasury (Room 42 III) on the 7th November, 1941," NA, T 161/1405.

[63] J. H. Burrows to Mr. Cleary, 15 Dec. 1941, NA, ED 11/269.

now nothing at all to prevent over 150 children from running on to the roads." The author of the letter added that his organization had asked the local authority to leave the railings but had been rebuffed.[64]

Writing to a colleague in November 1941, a Board of Education official named R. N. Heaton noted that "if we are to prevent the indiscriminate removal of railings necessary for the safety of the children or for the protection of school buildings or property the matter should be taken up at [a] high level." This was easier said than done. Because of the complicated, even Byzantine, layers of bureaucracy involved in the requisition of iron and steel, a great deal of confusion existed about who was in charge. Heaton explained that the Salvage Department of the Ministry of Supply had referred him to the Iron and Steel Control, which in turn suggested that he contact the Ministry of Works. After reaching an official there, he was advised to get in touch with the Raw Materials Department of the Ministry of Supply, the offices of which had been evacuated from London to Warwick Castle.[65]

The Ministry of Works, which had responsibility for removing railings, objected to giving the Board of Education authority to veto removals. Instead, it suggested that all parties should adopt a general policy, such as preserving only those railings that separated school grounds from busy roads. Reporting on a meeting that he had held with officials at the Ministry of Works on 5 December 1941, a Board of Education official remarked, "It is impossible to say how far the Ministry of Works and Buildings would be able to take a sympathetic line in dealing with school railings. They are being pressed very hard by the Steel Control to deliver up an amount of scrap which they consider they will not be able to collect."[66]

A short time later the Board of Education contacted local education authorities regarding the removal of railings from schools. "There is an urgent need," its letter noted, "for scrap iron and steel in order to maintain steel production for the war effort." Railings should be left standing, however, where they were needed to prevent accidents or to keep intruders from damaging school property. "Where it is considered that any scheduled railings should be retained, full particulars . . . should be sent to His Majesty's Inspector" of Schools.[67] Although the Board of Education failed

[64] Percival Laurence Leigh-Breese, letter to the editor, *Times* (London), 18 Dec. 1941, 5.
[65] R. N. Heaton to J. H. Burrows, 14 Nov. 1941, NA, ED 11/269.
[66] H. B. Jenkins, notes of a meeting at the Ministry of Works, 5 Dec. 1941, NA, ED 11/269.
[67] Board of Education, Administrative Memorandum No. 340, 19 Dec. 1941, NA, ED 11/269.

to obtain a blanket exemption against the removal of railings from school boundaries, their protests were not entirely in vain. On 29 January 1942, the Directorate of Demolition and Recovery (an agency within the Ministry of Works) ordered its staff to proceed carefully with school railings. "Before the work of removing railings is commenced in any area where school railings have been scheduled, the Local Education Authority must be asked whether they have made, or intend to make, representation in accordance with the Board of Education Administrative Memorandum No. 340." Only after a final decision was made were railings to be taken.[68]

In contrast to school officials, who almost never made aesthetic arguments in favor of railings preservation, many individuals who filed appeals did so on the basis of the artistic merit of their railings. In December 1941, T. N. Waldron, who belonged to the town council of Stratford-upon-Avon, complained, "I was surprised and shocked to learn that my entrance gates are scheduled to be taken and melted up as scrap for they are 'specimen' wrought iron gates of considerable beauty and I hereby enter a protest against them going as scrap."[69] When his appeal failed, Waldron responded that he was "not prepared to scrap them at the ridiculous price of 25/-[shillings] per ton. I am, therefore, having them removed and put on the scrapheap at the works where we shall obtain for them 55/-per ton which is the correct price for clean scrap."[70] A similar letter came from S. A. F. Mathison, who informed the town clerk that her gates and railings had been "specially designed at great cost for the purpose which they serve, and have considerable artistic merit."[71] The authorities responded with equal resolve to quash an appeal from Evelyn Nicholle, who wrote, "I should be glad if you could see your way clear to allow our Railings to remain." Her late father had created them, she explained, and ever since his death, "they have held for my mother a very great sentimental value."[72]

68 Directorate of Demolition and Recovery, memorandum, 29 Jan. 1942, NA, ED 11/269. On the formation and remit of the directorate, see Charles Mendel Kohan, *Works and Buildings*, History of the Second World War, United Kingdom Civil Series (London: HMSO, 1952), 221–2.

69 T. N. Waldron to P. C. Smart, 16 Dec. 1941, Shakespeare Centre Library and Archive, Stratford-upon-Avon, BRR 55/14/31/5/2.

70 T. N. Waldron to P. C. Smart, 10 Mar. 1942, Shakespeare Centre Library and Archive, Stratford-upon-Avon, BRR 55/14/31/5/2.

71 S. A. F. Mathison to town clerk, n.d. [Oct. 1941], Shakespeare Centre Library and Archive, Stratford-upon-Avon, BRR 55/14/31/5/2.

72 Evelyn Nicholle to town clerk, 28 Sept. 1941, Shakespeare Centre Library and Archive, Stratford-upon-Avon, BRR 55/14/31/5/2.

In March 1942, a member of the Walsall town council asserted that residents "were seething with discontent over the damage which was being done."[73] That same day a local newspaper published a letter from W. Everton, who complained that the program of railings removal was "un-English." In addition to asserting that the government had failed to make the case for why it needed to requisition private property, Everton claimed that the procedures it was following blended fascist brutality with socialist goals:

> On arriving home on Wednesday I found that the iron railings and coping in front of my house had been smashed in the best Nazi manner with crowbar and sledgehammer and flung into my front garden without my knowledge or consent. . . . The larger the house the greater the damage seems to be the case. The object appears to be to make us all feel the pinch with the maximum of annoyance, alarm and despondency.[74]

Not everyone shared this view, however. The loss of one's railings, argued Private A. Hitch, was a minor inconvenience compared to the sacrifices that many were making in wartime. "Mr. Everton," he wrote, "seemed extremely annoyed at the damage done to his garden when the much-needed scrap railings were removed without his consent. I was called up, like so many more, without my consent. I knock lumps off my hands which, pre-war, were needed for very fine work. My garden, of which I was so proud, lacks attention. I am away from the people I love." He concluded by noting, "There never will be equality of sacrifice, but we ought all to face our war trials cheerily."[75]

Such arguments failed to silence the critics of requisition. In December 1942, Erik Addyman, a resident of Harrogate, complained that the government's declaration that it would remove only "unnecessary" railings had naturally led the public to assume "that antiquities, high class workmanship, beautiful and artistic work would be spared." The reality was far different, he asserted. "In a few fortunate districts the work was under control of sympathetic people, [but] other districts were not so fortunate and in some districts the work was controlled by absolute vandals, whose one aim seems to have been to destroy everything of beauty or value." He alleged that to make matters worse, "the most underhand methods were sometimes adopted, owners being lulled to a sense of security that their interests would be respected and then when it was thought they were

[73] "'Attack' Starts on Walsall's Railings," *Walsall Times*, 21 Mar. 1942.
[74] W. Everton, letter to the editor, *Walsall Observer*, 21 Mar. 1942.
[75] A. Hitch, letter to the editor, *Walsall Observer*, 4 Apr. 1942.

away from home the beautiful gates and railings would be stolen and destroyed. . . . There are numerous cases of property worth £100 or more being absolutely wrecked for the sake of a few hundred weight of iron not worth ten shillings." It was difficult to imagine, he asserted, how anyone treated this way would give "a single further penny to government loans when there is the slightest chance of contributing by so doing to the Salaries of the destroying vandals." Addyman objected not only to the monetary loss and destruction of beauty but also to the blatant disregard that officials had shown toward the values of "fairness and fair play." In closing his letter, he declared, "We all want to help to win this war, but this ruthless raiding of the Englishman's home which was once his castle is the wrong way to go about it."[76] Referring to this criticism, one civil servant sardonically remarked, "It rather looks as if one or two of the cases mentioned may have been somewhat unwisely handled."[77]

Instead of writing letters of complaint, some people took more direct action. One woman was so upset when she learned that her railings were being requisitioned that she scared away the "hooligans" sent to remove them by threatening to let loose her four St. Bernards on them. When their foreman learned of her threat, he authorized his men to use their oxyacetylene torches to defend themselves. Ready to do battle, the men returned, only to find that the railings had disappeared. Their owner had hidden them in her back garden. When confronted, she replied that she would only give them up if the government promised to replace them after the war. This act of resistance landed her in court, where she was fined £20.[78]

R. H. North, a resident of the small town of Hessle in the East Riding of Yorkshire, also took matters into his own hands. When he lost his appeal to retain his gates, he removed them and refused to tell the authorities where they were. In response, the Ministry of Works warned him that if he did not surrender the gates, he would be charged with "an offence under the Defence Regulations."[79] This prompted North to contact the chancellor of the exchequer, Sir Kingsley Wood. "As a protest against my persecution by the Ministry of Works + Buildings," he wrote, "I beg to send herewith 8 Savings Certificate books and one 3% War Stock certificate, the whole of my family's savings since the war began

[76] Letter from Erik T. Addyman, 8 Dec. 1942 (copy), NA, T 161/1405.
[77] H. E. C. Gatliff to R. M. Hunter, 22 Jan. 1943, NA, T 161/1405.
[78] "The Removal of Iron Railings," *Justice of the Peace and Local Government Review*, 14 Aug. 1943, 392.
[79] H. L. Raybould to H. E. C. Gatliff, 11 Nov. 1942, NA, T 161/1405.

being in these securities. Will you please cause me to be repaid the whole of my 500 Certificates and the War Stock in order to furnish me with funds to defend myself against the Ministry concerned."[80]

Commenting on this petition, a Treasury official recommended to his colleagues that it was "absolutely essential" for the government to "take a firm line in cases of this kind; the problem will arise in a far more serious way over motor tyres and even more over non-ferrous metal articles if, in the last resort it proves necessary, which we devoutly hope it won't, to requisition domestic treasures." Alleging that the author of the complaint exhibited "signs of persecution mania," the Treasury official H. E. C. Gatliff quipped that "the writer may go mad before anything happens, but I am not very hopeful of this as such cases as a rule develop slowly and anyhow he may be merely suffering from normal irritation."[81]

In January 1942, Major-General Sir Alfred Knox grilled Harold Macmillan, who at that time was parliamentary secretary to the Ministry of Supply, about the procedures that were being followed in the removal of railings from private property. Knox alleged that the scheduling of railings for removal was being conducted "largely by voluntary women workers who are not really competent to carry out the work." Macmillan declined to address this issue, but he did respond to Knox's question about why owners were not told in advance that their railings were going to be taken. According to the future Conservative prime minister, it was simply too difficult to track down everyone who had an interest in railings scheduled for removal.[82]

Members of parliament also complained that severe damage was being caused to stone and brickwork as railings were being removed. George Hicks, parliamentary secretary to the Ministry of Works, replied that "some incidental damage to property is inevitable," but that its contractors would be responsible for repairing any damage caused by negligence on their part.[83] In the case of damage that could not have been avoided, the Ministry of Works adopted a policy of paying for "limited repair."[84] Much of the damage that took place resulted from a shortage of appropriate equipment. Because not enough oxyacetylene torches or hacksaws existed and because of the urgent need for scrap, workers often used

[80] R. H. North to Kingsley Wood, 6 Nov. 1942, NA, T 161/1405.
[81] H. E. C. Gatliff to H. B. Usher, 12 Nov. 1942, NA, T 161/1405.
[82] "Removal of Railings," *Municipal Review*, Feb. 1942, 22.
[83] *Parliamentary Debates*, Commons, 4 Feb. 1942, vol. 377, col. 1197W.
[84] Ministry of Works and Planning, memorandum, 15 Oct. 1942, NA, AO 30/32.

heavy hammers to break railings free from their stone settings.[85] The ensuing damage was in many cases extreme, and it prompted the Ministry of Works to inform demolition contractors that "complaint has been laid in the highest quarters" about the manner with which some firms were proceeding. According to these reports, workmen had not only caused considerable damage to stonework, but had obstructed walkways with sections of dismantled railings in places where people tripped over them in the blackout. Furthermore, members of the public had lodged "many complaints of bad language and incivility on the part of Contractors' employees."[86]

Glaring inequalities existed from one place to another in terms of the generosity of repairs. In the twelve months from April 1942 to March 1943, the government spent over half a million pounds "making good" damage caused by the removal of railings in London, but only £100,000 for this purpose throughout the rest of the country. For every ton of railings that it collected in London, the government spent nearly £4 on repairs, a rate that was twelve times greater than what it paid per ton in Newcastle, and fifty-seven times what it paid per ton in Liverpool.[87] The most likely explanation for this disparity is the greater wealth and political influence of those who owned land in the capital. This group included not only powerful private citizens, but also government departments, the London County Council, and the crown.

By the spring of 1942, in the face of growing public opposition to requisition and mounting costs to repair damage to stone and brickwork, the Treasury official H. E. C. Gatliff suggested that perhaps the need for scrap was no longer "quite so urgent as it had been" the previous autumn. He added that the Ministry of Works had "been pretty rigid about collecting nice gates etc. but now apparently the Minister is showing signs of getting cold feet. . . . From the financial point of view this may be as well."[88] In contrast to Gatliff, whose job it was to consider railings from a fiscal perspective, officials in the Ministry of Works remained convinced that other factors should be given precedence. America's entry into the war, suggested one official, meant that it would be unlikely to send any scrap to the United Kingdom for the foreseeable future. As a result, he

[85] George Hicks to George Schuster, 28 Mar. 1942, Walsall Local History Centre, 235/4.
[86] Demolition and Recovery Officer, Area No. 9 to local authorities, copy, [Mar. 1942], Walsall Local History Centre, 235/4.
[87] Expenses of Collection and Recovery of Scrap Metals, n.d. [1943?], NA, AO 30/32.
[88] H. E. C. Gatliff to J. S. Bromley, 14 Apr. 1942, NA, T 161/1405.

warned, "The nation is in desperate straits, and every ounce is wanted. I can assure you that if this were not so we should not allow ourselves to be put to the trouble and to incur the odium involved in taking the ordinary citizen's railings."[89]

Pressured to salvage every possible bit of material, officials looked anew at ironwork that they had previously left standing. In a 1943 letter to the editor, a Yorkshire farmer complained that despite the government's promise to leave in place railings necessary to keep livestock from straying, laborers employed by the Ministry of Works had dismantled nearly a hundred feet of fencing, "thus throwing 47 acres of grazing land open to the highway." The writer added that government agents seemed to be interested only in "light stuff, easy to handle, because, only a few yards farther along, the heavy portion of a 1914–18 German gun still lies unclaimed by the side of the road, though the lighter parts were removed two years ago; while the total weight of my 30 yards of railings is about 1 cwt. [112 pounds], and its value as scrap but a few shillings."[90] As such complaints grew louder, those closest to the railings removal program became increasingly strident in their defense of the unpopular policy. George Hicks, parliamentary secretary to the Ministry of Works, declared in 1943, "I have taken railings from 2,500,000 houses as raw material for munitions. If you get through this war losing nothing more than railings, you can be jolly thankful."[91]

During the first year of requisition, owners requested compensation in 300,000 cases, which amounted to 15 percent of all railings taken.[92] Many missed the chance to file a claim, for the government required owners to submit a petition no later than six months after their railings were taken, Because so many people were away from home in wartime, this often proved impossible. The government hoped that the short deadline would not only minimize the number of claims filed, but also the size of each claim. As H. E. C. Gatliff explained to a Treasury colleague in 1942, owners would "stand a much better chance of something approaching reinstatement value after the war when the public mind is less ready to condone our running away with such treasures, practically without paying for them."[93] Writing to another official that same

[89] E. J. R. Edwards to H. B. Jenkins, 2 Apr. 1942, NA, ED 11/269.
[90] Osbert Sitwell, letter to the editor, *Times* (London), 26 Mar. 1943, 5.
[91] "Scrap Metal Shortage," *Public Cleansing and Salvage*, Feb. 1943, 212.
[92] *Parliamentary Debates*, Commons, 30 Sept. 1942, vol. 383, cols. 764–5.
[93] H. E. C. Gatliff to R. M. Hunter, 25 Sept. 1942, NA, T 162/748.

year, Gatliff observed, "It seems to be generally agreed that the steel is worth in use at least £10 a ton, and while [the Ministry of] Works will try and settle [compensation claims] at 25/-a ton it would be worth paying a lot more for the stuff. The risk is, of course, that this may lead to some undeserving people getting an unreasonable amount, and that might have awkward reactions. The Parliamentary Secretary has said that Works must throw up to the Tribunal anything they can't settle within £10 a ton so as to protect them from political criticism."[94]

In a further attempt to limit the size of the payments it would have to make, officials in the Ministry of Works obtained a slight amendment to the legal provisions that governed requisition. The new rule, Defence Regulation 50B, stipulated that owners could no longer claim compensation for a decline in the value to their land, but only for the loss of fixtures. Despite this change, explained an official in the Ministry of Works, "It is by no means certain that if cases go before the Tribunal the amount awarded will be based on the value of the fixtures as scrap. We should argue that as there is no market for railings at the moment the price which could be obtained on a sale would be the scrap metal value, but the Tribunal might not accept that view. If the Tribunal ruled that the compensation should be the price which could, in ordinary circumstances, be obtained for railings, the figure would exceed £10 per ton for the cheapest types."[95] Commenting on this, Gatliff noted, "Railings don't cost much to collect, but may cost a lot in compensation if either they are particularly good railings or they were not so unnecessary as [the Ministry of] Works thought."[96]

In January 1942, the National Federation of Property Owners, based in London, and a similar group in Glasgow, petitioned the chancellor of the exchequer, Sir Kingsley Wood, to allow owners of requisitioned railings to be compensated under the War Damage Act that had become law the previous year. Both groups expressed their support for the government's efforts to take railings, but they maintained that it had a responsibility to pay for the cost of replacing railings in cases where it was "necessary or desirable to do so." They argued that the government should give just as much consideration to owners whose property was damaged as a result of government policy as it did to those who sustained losses as a direct result of enemy action.[97]

[94] H. E. C. Gatliff to H. B. Usher, 29 June 1942, NA, T 161/1405.

[95] H. L. Raybould to H. E. C. Gatliff, 9 June 1942, NA, T 161/1405.

[96] H. E. C. Gatliff to H. M. Young, 18 June 1942, NA, T 161/1405.

[97] T. Simpson Pedler to Kingsley Wood, 27 Jan. 1942, NA, T 162/748; R. Murray MacGregor to Kingsley Wood, 29 Jan. 1942, NA, T 162/748; "Search for Scrap," *Times* (London), 31 Jan. 1942, 2.

Although the British government refused throughout the war to pay for the cost of replacing railings, it eventually agreed to compensate owners on the basis of their "reasonable second-hand value." This change increased the compensation rate at least tenfold: on this basis, even the most ordinary of railings were valued at £13 to £15 a ton.[98] Yet the new policy applied only to owners who had pending claims, not those who had failed to request compensation within the required six-month window or those who had accepted compensation at the original rate of 25 shillings per ton. At the same time that discontent was brewing over the requisition of railings, a horrific event occurred that renewed concerns about whether the government had neglected public safety in its rush to obtain scrap metal.

UNINTENDED CONSEQUENCES

The largest single loss of life from a single incident in Britain during the Second World War came not from bomb blasts or fire, but because of a misstep in a crowded stairwell. The tragedy occurred at the entrance to the Bethnal Green Underground Station, which, like many tube stations, served as a shelter during bombing raids. When the air raid sirens sounded shortly after 8 P.M. on 3 March 1943, many Londoners took notice. The radio and newspapers had been full of reports about the increasingly heavy Allied bombardment of Berlin, and many Londoners expected Hitler to retaliate with a massive attack on their city. In the metropolitan borough of Bethnal Green, loud concussions echoed through the streets. Though caused by the firing of anti-aircraft guns, some residents thought that the sound came from exploding bombs, and they rushed toward the tube shelter. Rain had fallen a short time earlier, and the steps that descended from street level were slippery. The stairwell was also extremely dark; the only light came from a solitary low-wattage bulb. When a woman carrying a baby tripped at the bottom of the stairs, a deadly chain reaction occurred in which hundreds of people fell forward, trapping and suffocating those beneath them. The disaster claimed the lives of 173 men, women, and children.

Over a year earlier, workmen employed in the scrap-metal campaign had taken the iron railings that surrounded the station entrance and replaced them with wooden hoardings. The railings had performed more

[98] R. M. Hunter to H. E. C. Gatliff, 10 May 1944, NA, T 162/748.

than just a decorative purpose, for they had separated the surface level from the stairwell and restricted the number of people who could enter it at any one time. Soon after the removal of these railings, a committee of the town council issued a prophetic warning to officials with the London Civil Defence Region (LCDR): "There is a grave possibility that, on a sudden renewal of heavy enemy air attack, there would be an extremely heavy flow of persons seeking safety in the Tube Shelter, and that the pressure of such a crowd of people would cause the wooden structure to collapse, and a large number would be precipitated down the staircase." To reduce the likelihood of such an accident, borough officials asked the LCDR to approve the expenditure of £88 "for the removal of the hoarding and the construction of brick wall and piers, re-erection of railings and hanging of gates at the entrance to the Tube Shelter."[99] Despite repeated pleas from the borough, regional officials refused to approve the work. In response to this opposition, the councilors warned that they could not "accept any responsibility for the consequences which might ensue from the lack of adequate protection for the entrance to the Shelter." After reviewing the matter, the technical adviser to the LCDR declared, "I am still of the opinion that it would be a waste of money to build up a wall round the steps to prevent the crowd from forcing their way into the shelter." The commissioners of the LCDR agreed, and they refused to take action.[100]

The inquiry that followed the tragedy largely blamed the victims and concluded – apparently on the basis that the hoardings had remained intact – that the removal of railings had played no role in the tragedy. Yet as the sociologist Charles Perrow has argued, accidents are usually the result of multiple interacting factors.[101] Slippery surfaces, poor visibility, concussive sounds, and the expectation of a major attack each contributed to the tragedy, and the removal of the station's railings may have done so as well. Although the official report did not consider whether the modification to the station entrance increased the number of people who might exert pressure on others who were in the staircase, this seems

[99] S. P. Ferdinando to chief administrative officer, London Civil Defence Region, 20 Aug. 1941, NA, HO 207/997.

[100] Great Britain, *Report on an Inquiry into the Accident at the Bethnal Green Tube Station Shelter on the 3rd March, 1943* (London: HMSO, 1945), esp. 29–33. This report has been reprinted as *Tragedy at Bethnal Green* (London: Stationery Office, 1999).

[101] Charles B. Perrow, *Normal Accidents: Living with High Risk Technologies*, rev. ed. (Princeton, N.J.: Princeton University Press, 1999).

plausible. If this in fact occurred, then the removal of the railings must be included among the causes of this accident.

Speaking in Parliament a few months after the Bethnal Green tragedy, Lord Hemingford suggested that the government's quest for railings had "caused a considerable amount of obloquy to fall upon His Majesty's Government. It has been felt that an injustice had been done to a very large number of usually uncomplaining and patriotic people. . . . These regulations are one of the results of hasty legislation in the early days of the war, for which every one of us who was a member of either House of Parliament at that time must take his share of responsibility." Wyndham Portal, now minister of works, replied, "Nobody deplores more than I or His Majesty's Government do, having to take anything away from people, but I should like to take your Lordships' minds back to the time when we started to remove railings and gates. It was in August or September, 1941, that we had to stop importing scrap from America, in order to save shipping space and also because the Americans wanted scrap for themselves." According to Portal, railings had provided a vital supply of scrap metal at a critical moment, "because we had no scrap available at that time to compare with what we had been bringing over from America."[102]

Requisition of railings continued until 31 August 1944, when the Ministry of Works official G. M. Carter ordered workers to stop removing the small number of railings that remained.[103] By this point, according to one estimate, officials had taken 582,882 tons of railings for the war effort, as well as nearly an equal amount of blitz scrap. Tram rails "cleared and purchased" amounted to 114,461 tons.[104] One factor that may have influenced the government's decision to stop collecting railings was the heavy bombardment that Britain faced from German flying bombs and rockets during the summer and fall of 1944. These weapons caused enormous damage and loss of life. Between mid-June and late

[102] Parliamentary Debates, Lords, 13 July 1943, vol. 128, cols. 437-43; also quoted in "Requisitioned Railings," *Public Cleansing and Salvage*, Aug. 1943, 424.

[103] Directorate of Emergency Works and Recovery, Instruction No. 193, 31 Aug. 1944, NA, AO 30/32.

[104] British Iron and Steel Corporation (Scrap), Report to Directors, Board Meeting, 28 Nov. 1945, 2–3, NA, AO 30/32. Emphasis in original. Official sources cover a range of periods and provide somewhat varying estimates of the total cost of collection and the quantities obtained. For other estimates of the quantities of ferrous materials collected, see Expenses of Collection and Recovery of Scrap Metals, n.d. [1943?], NA, AO 30/32; *Parliamentary Debates*, Commons, 27 May 1943, vol. 389, cols. 1733–4; and R. Holt, "Railings," 6 Apr. 1944, NA, WORK 22/430.

September 1944, V1 flying bombs and V2 rockets destroyed 25,000 buildings and caused serious damage to 130,000 others.[105] The Ministry of Works quickly became overwhelmed by the need to make emergency repairs to roads, public buildings, and houses. Under the circumstances, the thought of using the country's finite pool of labor power to dig up old tram lines or harvest railings made little sense, particularly because these new attacks produced such vast new quantities of blitz scrap.

More than any other form of salvage in Britain during the Second World War, the requisition of railings affected the appearance of nearly every city and town in the kingdom. More than four decades after the railings were removed, one resident of London observed that "the scars still remain today outside many great buildings and squares." As an example, he pointed to "the row of rusting stumps outside the Royal Geographical Society, where the railings were crudely broken off."[106] The campaign to collect and melt down Britain's iron railings exemplified the shift toward total war. It was also one of the most controversial and legally dubious emergency measures that the government took in wartime, and one of the most misunderstood. It raised issues of class conflict, history, individual rights versus collective security, ends versus means, the influence of the United States on British domestic policy, and fairness. Although no consensus is likely to emerge about whether the government ought to have proceeded as it did, there can no longer be any doubt that railings indeed helped to feed the war machine.

[105] Charles Mendel Kohan, *Works and Buildings*, History of the Second World War, United Kingdom Civil Series (London: HMSO, 1952), 222–5.

[106] Roy Miles, letter to the editor, *Times* (London), 23 Jan. 1985, 15.

10

Victory and Postwar

> Some . . . regard salvage as a war-time fad, but make no mistake, when we have won this war it will be necessary to continue salvage of material as a more economical and scientific method of disposing of refuse than we have hitherto employed.
>
> –*Municipal Journal & Local Government Administrator*, 1940[1]

During the final two years of the war the need for raw materials remained as great as ever. Britain's mines, forests, and farms could not supply enough ore, wood pulp, and food to fulfill wartime requirements, and neither could imports. Recycling helped to bridge that gap. At the same time that the government sought to maintain or even boost salvage collections, it faced mounting challenges. Many of the country's saucepans, mementos, railings, books, and papers had already been recycled for the war effort, and the government faced growing criticism for the way it had gone about collecting them. Some critics, seeing piles of unutilized scrap metal, wondered whether the entire salvage effort was simply a charade. Others were simply exhausted.

As the country's munitions production reached its peak in 1943, over forty thousand WVS members were involved in salvage work. In addition to assisting in educational and publicity efforts, women went door-to-door to collect household scrap, peeled wrappers off food containers, and sorted books, paper, metal, and animal bones.[2] In February 1943, the

[1] "The War on Waste," *Municipal Journal & Local Government Administrator*, 23 Aug. 1940, 1100.

[2] "W.V.S. and Salvage," WVS *Bulletin*, June 1946, 6; Women's Voluntary Service for Civil Defence, *Report on 25 Years Work, 1938–1963* (London: HMSO, 1963), 74.

WVS *Bulletin* declared that "the word 'Salvage' is now written across every W.V.S. member's heart, for all over the country members canvass, collect, sort and bale."[3]

Under the heading "Another Salvage Drive," the *Times* reported in June 1943 that a campaign "on a scale hitherto not attempted" would soon take place throughout Britain. Speaking in Trafalgar Square on the final day of the drive, Lady Mountbatten asserted that the collection of salvage was "probably the most important of all the war jobs" that existed.[4] After years of war, however, many overworked adults had become weary of calls to hand over yet more scrap. As a result, the government looked increasingly to children. More than 300,000 child "commandos" volunteered in London alone to take part in the June 1943 book drive, and officials expressed the hope that they would collect five million books in the capital. "Badges are being provided, and promotions will be made according to the number of books individually collected."[5] The rank of general went to those who brought in 150 books; to become a field marshal, a boy or girl had to collect 250 volumes. Some children managed to gather hundreds more than this.[6] In the West Midlands town of Walsall a pair of twins, aged eight, brought in over 1,700 books between them.[7]

CONSCIENCE

Apathy and exhaustion were not the only reasons that some people failed to recycle. In May 1943, more than a year after the government made it illegal to discard or destroy paper, a salvage collector in St. Leonards-on-Sea, a town adjacent to Hastings in southern England, discovered paper in two dustbins that belonged to Muriel and Donald Jackson. The official knocked on the door and informed Mrs. Jackson of his discovery, and she listened without saying a word. Perhaps because this response was so unusual, town officials decided to keep close watch on the Jacksons' bins. Each time they checked, they found paper in the couple's rubbish. In early June, after a salvage worker confronted Muriel Jackson a third time and told her that he would have to report her, she said, "I absolutely refuse to

[3] "Salvage News," WVS *Bulletin*, Feb. 1943, 2.

[4] "Greatest War Work of Women," *Sunday Times*, 20 June 1943, 6.

[5] "Another Salvage Drive," *Times* (London), 2 June 1943, 8.

[6] "Salvage Work: No 'Glamour' But . . ." *Public Cleansing and Salvage*, Aug. 1944, 494.

[7] "Walsall's Book Drive," *Walsall Observer*, 13 Nov. 1943.

save anything for making munitions for killing people." When Mr. Jackson arrived home, he challenged the inspector to report the violation.

Two weeks later, a municipal employee again discovered paper in the Jacksons' dustbin, as well as rags and books. Wartime regulations expressly forbade each of these items from being discarded. This time, the police came to investigate. When the officers demanded to know why the couple had violated the salvage regulations, Muriel Jackson answered, "We are trying to make it clear to the dustmen that we do not intend to keep the refuse separate." They explained that they were Quakers and that they opposed war. If they contributed paper, rags, and books to be converted into weapons, they would violate their religious commitment to nonviolence.[8] For the same reason they refused to recycle their rubbish, the Jacksons objected to the government's plan to requisition their railings. When the workmen assigned to sever their railings arrived at the Jacksons' house, the railings were already gone. Donald Jackson had removed them himself and buried them in the couple's garden.[9]

Wartime salvage regulations contained no exemptions for conscientious objectors, so local officials decided to bring the Jacksons to trial. The deputy town clerk, Stephen King, served as the prosecutor. He charged Muriel with two counts of mixing waste paper with refuse, and one of doing the same with rags. He accused Donald Jackson on a single count of aiding and abetting his wife in her criminal activities. This was not Donald Jackson's first brush with the law. Shortly after the outbreak of the war he had been imprisoned for refusing to join the military, but he was released shortly thereafter when an appeals board granted him conscientious objector status. In the spring of 1940, at great personal risk, Jackson crossed the English Channel in a small boat to help evacuate Allied soldiers from Dunkirk. Instead of disputing the charges against them, the Jacksons admitted that they had broken the law. Their defense rested entirely on their good character and their religious objection to assisting, however indirectly, in the deaths of others.

Little sympathy existed in wartime Britain for "conchies," as they were often scornfully called. If a man declared that he was a conscientious objector, people often expressed doubts about his masculinity and his

[8] "Salvage Mixed with Refuse: 'Conscience' Plea by Local Couple," *Hastings & St. Leonards Observer*, 21 Aug. 1943, 1; "Hastings Prosecution," *Public Cleansing and Salvage*, Nov. 1943, 128.

[9] Author's interview with Evelyn Stanley Jackson, St. Leonards-on-Sea, 27 July 2009. I am very grateful to Ms. Jackson and to her son, Mr. Philip W. Holland, for their generosity.

patriotism. Attitudes were particularly harsh in coastal areas such as Hastings and St. Leonards, where residents faced repeated attacks from German planes and the possibility of invasion from the sea. Two months before the authorities observed the Jacksons' failure to recycle, bombs killed thirty-eight residents of the community and destroyed a large number of houses, including some on the street where the Jacksons lived.[10] As an expression of outrage against pacifists, in August 1940 the Hastings town council had voted by a large margin to fire two town employees because they were conscientious objectors. One of their few defenders on the council was Alderman Frederick Morgan, a local solicitor, who argued that because British law recognized the right to conscientious objection, it was wrong for the council to punish its employees for holding such opinions.[11]

When the Jacksons went to trial, they needed someone who was not afraid to defend their acts of civil disobedience in a time of overwhelming support for the war. Not surprisingly, perhaps, they approached the Morgan law firm. But instead of retaining Frederick Morgan, they chose his daughter Dorothy, who shared the practice with him. Eighteen years earlier, at the age of only twenty-four, Dorothy Morgan had become the first female solicitor to appear in court in Hastings. Reporting on that milestone when it occurred, the local newspaper had observed condescendingly that "amongst the papers in front of Miss Morgan was her black handbag, a delightfully feminine note amongst the official looking documents. She stated her case in a clear voice without a trace of nervousness."[12]

In the courtroom Morgan pursued a curious strategy, one that perhaps derived from her calculation that the court would ultimately consider a husband culpable for both his and his wife's actions. Morgan said very little about Muriel Jackson except that she shared her husband's views – a statement that left room for ambiguity about whether she did so out of conviction or duty – and asserted that she "had been perfectly frank throughout" the investigation. Morgan moved on to praise Donald Jackson for his community service work, noting that he belonged to a first-aid team and served as a firewatcher who on several occasions "showed the utmost bravery and disregard for his own life." Donald Jackson, she explained, "does all he can to save life, but in this matter he

[10] Nathan Dylan Goodwin, *Hastings at War, 1939–1945* (Chichester: Phillimore, 2005), 74.

[11] "Council Sack 'Conchies,'" *Hastings & St. Leonards Observer*, 3 Aug. 1940, 4.

[12] "Lady Solicitor," *Hastings & St. Leonards Observer*, 14 Feb. 1925, 8.

is following the dictates of his conscience and he thinks it is wrong to keep waste materials so that they can be turned into weapons for taking life."[13]

The judge, or chairman, in the case was Arthur Blackburn, a coal merchant and alderman who had served four terms as mayor of Hastings. Three years earlier, he had pledged £1,000 of his own money – a sum that was then several times greater than the average annual salary – to a local effort to sponsor the purchase of a Spitfire fighter plane.[14] The verdict was something of a compromise. Although Chairman Blackburn ruled that the Jacksons had broken the law, he did not wish to make martyrs of them. Instead of handing down an immediate punishment, he bound them over (sentenced them to probation) for a period of twelve months. "If you separate the material properly during that period you will hear nothing more about it, but if you act as you have been doing you will be brought before the court for judgment."[15]

This was a lenient decision, for the magistrates could have sentenced both Donald and Muriel Jackson to prison, imposed a large fine, or both. Instead of quietly accepting the conditions of their release, however, Donald Jackson wrote an outspoken letter to the editor of the local paper:

> Sir, – In your report in last week's issue of your paper, regarding the prosecution against us for mixing salvage with our refuse, I would like to point out a wrong impression has been given by this résumé.
>
> While we do not wish wantonly to violate the law, we cannot obey the law of our country when it conflicts with the dictates of our conscience, so that while we have agreed to refrain from mixing salvage with our refuse . . . we have not agreed to save any salvage for munitions, and wish to make this clear.

Jackson ended his letter by commending the magistrates who heard the case, "who did all they could to understand our position."[16]

In sharp contrast to the Jacksons' refusal to contribute materially to the war effort, the sculptor Clare Sheridan embraced the opportunity. In January 1943, several months before the Jacksons attracted the attention of the authorities in their town, Sheridan boasted to her first cousin, Winston Churchill, that she had recently "contributed a couple of bronze heads to be melted down – (one of which is a head of Rakovski who was

[13] "Salvage Mixed with Refuse," 1.

[14] "'Observer' Launches 'Spitfire' Fund for Hastings," *Hastings & St. Leonards Observer*, 24 Aug. 1940, 4.

[15] "Salvage Mixed with Refuse," 1.

[16] Donald B. Stanley Jackson, letter to the editor, *Hastings & St. Leonards Observer*, 28 Aug. 1943, 4.

Soviet Ambassador here!)"[17] Sheridan had met Christian Rakovsky, as his name is generally transliterated, on a visit to Russia in 1920. There she had created busts of many leading Bolsheviks, including Lenin, Trotsky, Kamenev, and Rakovsky – all of whom, with the exception of Lenin, would eventually be killed at Stalin's behest. Rakovsky, who had represented the USSR at the Court of St. James in the mid-1920s, eventually found himself on Stalin's long list of enemies. On 11 September 1941, on direct orders from Stalin, secret police from the NKVD (a forerunner to the KGB) marched Rakovsky and more than a hundred other prominent political prisoners (including Olga Kameneva, who was Trotsky's sister and Lev Kamenev's widow) from Oryol Prison to the adjacent Medvedev Forest and shot them.[18] Sheridan's casual destruction of Rakovsky's bust in the wake of his death bears a disquieting similarity to the erasure of disgraced Communists from official photographs and texts that routinely occurred under Stalin's rule.[19] Wartime salvage, as the cases of Donald and Muriel Jackson and of Clare Sheridan make clear, raised important ethical questions about whether and in what ways civilians should participate in warfare. Salvage also raised vital questions about the proper balance between civil liberties and state power in a time of national crisis.

On occasion, the British press collaborated with the government's efforts to silence criticism of its salvage policies. In July 1943 a London landlord named L. Pickrell wrote a short letter to the editor of the *Times* about his recent experiences with requisition. After losing an appeal against the seizure of his custom-made wrought iron gate, he demanded compensation for the cost of replacing it. The Ministry of Works informed him that the most he could claim was its value as scrap metal. Pickrell, enraged, alleged that the government was engaged in "legalised robbery."[20] Instead of publishing this complaint outright, the news editor of the *Times* forwarded the critique to the Ministry of Works for comment.[21] The ministry took the matter extremely seriously, and three

[17] Clare Sheridan to Winston Churchill, 7 Jan. 1943, Churchill Papers, Churchill Archives Centre, Churchill College, Cambridge, CHAR 1/374/2.

[18] Robert Conquest, *The Great Terror: A Reassessment* (New York: Oxford University Press, 2007), 456.

[19] For numerous visual examples of these erasures, see David King, *The Commissar Vanishes: The Falsification of Photographs and Art in Stalin's Russia* (New York: Metropolitan Books, 1997).

[20] L. Pickrell to the editor of the *Times* (London), 14 July 1943 (copy; never published), NA, WORK 22/430.

[21] Alex Pitt Robbins to Hugh Beaver, 19 Aug. 1943, NA, WORK 22/430.

officials, including the minister of works himself, visited the editor. They asked him not to publish the letter, and he complied.[22] On numerous other occasions, of course, the *Times* did publish letters from people who criticized railings requisition, but they generally used much less strident language than this unpublished letter.

TIN CANS

One of the biggest headaches of wartime salvage, at least from the perspective of local authorities in Britain, involved tin cans. People dutifully saved them, and the Ministry of Supply required local authorities to collect them, but few processors had the equipment necessary to strip the tin from the cans so their steel could be used. Also, their bulk made them difficult to store and to transport, and if they were not cleaned carefully, they would quickly emit foul odors and attract vermin. As a result, the government insisted that they be flattened or crushed into bales before being transported from collection dumps to factories. Few localities possessed the machines needed to do this. As a consequence, many salvage dumps accumulated large and unsightly piles of old cans. Even when cans were crushed and baled, local authorities often found it difficult to attract buyers for them. Given the choice between hauling away cans or removing more valuable materials, such as paper, bones, or old implements made of metal, scrap dealers and ministry employees usually chose the latter.[23]

Despite the long-standing complaints against them, many village dumps remained eyesores for years. As one official commented in 1942, their persistence came to symbolize inefficiency and waste: exactly the opposite of what the government hoped to achieve. As a result, the difficulties involving tin cans tainted people's faith in salvage as a whole, "with the repercussions one would naturally expect when we come along to take their house railings!"[24] Local officials, especially in rural areas, were extremely dissatisfied about the large dumps of cans and household odds and ends that they had collected but were unable to sell. In October 1943, Kent held the dubious distinction of the county with the largest quantity of

[22] Alex Pitt Robbins to Hugh Beaver, 20 Aug. 1943, NA, WORK 22/430.
[23] Ministry of Supply, Salvage Department, District No. 3, notes of a meeting in Birmingham, 29 Apr. 1942, 4, Shakespeare Centre Library and Archive, Stratford-upon-Avon, BRR 55/14/31/4/1.
[24] E. J. R. Edwards to R. H. Charles, 23 Apr. 1942, NA, ED 138/64.

unflattened tin cans – a stock of 1,101 tons. This represented approximately sixteen and a half million cans. Other counties with more than a million pounds of cans in the unflattened state were Essex (613.5 tons), Lancashire (529.5 tons), Norfolk (603.5 tons), Surrey (719 tons), Wiltshire (701.5 tons), and Yorkshire (782.5 tons). In Britain as a whole, local authorities possessed 11,385 tons of unflattened tin cans, or approximately 170 million cans.[25]

Prior to the decision in late 1943 of the Ministry of Supply to scale back the collection of "light ferrous scrap," 1,258 local authorities in England, Scotland, and Wales, along with seventeen in Northern Ireland, were required to operate collection programs for tin cans. In November 1943, the Ministry of Supply decided that collections should continue only in the 292 communities (two of which were in Northern Ireland) that possessed metal baling or flattening equipment.[26] Explaining the shift in policy, the ministry noted that

> the success of the Salvage Campaign and the efforts of Local Authorities has [*sic*] been such as to increase the supply of light scrap beyond requirements for consumption, which are limited by the uses to which this type of scrap can be put. The increased supply in the past has been of great help to the Steel Industry in augmenting our supplies of pig iron, but in present conditions it is desirable to reduce the quantity of this material now being collected and to devote the energy thus released to the collection of other forms of salvage, e.g. waste paper, textiles, household bones etc.

Materials were not the only thing that needed conserving, the circular explained. "Economy in the use of labour, road and rail transport is even more essential than hitherto."[27]

Addressing his fellow MPs on 30 November 1943, Charles Peat, parliamentary secretary to the Ministry of Supply, announced that the collection of cans had reached "a point exceeding the capacity of the industry to consume." Peat promised that the government would clear (and pay for) all of the accumulations that existed at the end of 1943. He declared that this decision did "not imply any change in the demand for other types of ferrous scrap suitable for direct use in the steel furnaces. The need for this is greater than ever. . . . The fact that we have achieved

[25] Ministry of Supply, Salvage and Recovery (L.A.), "Stocks of Salvage Materials Held by Local Authorities as of 31st October 1943," 1943, Shakespeare Centre Library and Archive, Stratford-upon-Avon, BRR 55/14/31/4/6.

[26] J. C. Dawes, minute on light ferrous scrap, 30 Nov. 1943, NA, BT 258/406.

[27] Ministry of Supply, Salvage Circular 107 (Nov. 1943), Walsall Local History Centre, 235/3.

so satisfactory a position on this one item of salvage does not reflect any lessening of the importance of salvage generally."[28]

People reacted to the announcement that they should no longer recycle tin cans with a mixture of confusion and frustration. Many wondered why the government had asked them to go to the trouble of cleaning, saving, sorting, and transporting their household cans. The most probable answer is that tin cans were such an obvious source of metal that the government had to make a point of collecting them to demonstrate to its own people and its American allies that it took salvage seriously. The great irony is that discarded cans, which it seemed so prudent to recycle, were not actually utilized, while railings, which still served a useful purpose, were from a strategic perspective more valuable as scrap.

The sight of large piles of metal cans lying in the open for months or years at a time sparked cynicism toward all salvage.[29] Because both railings and cans consisted of metal that the Ministry of Supply had declared to be vital to the war effort, the ministry's glaring failure to recover metal from cans naturally led many people to assume that the same was true of railings. This impression could not have been further from the truth, for railings provided a much more useful source of metal, with far lower labor, transportation, and processing costs, than did tin cans. The primary reason why the government pursued the farce of collecting tin cans that it could not actually use was to facilitate the collection of railings that it desperately needed. Paradoxically, although the (not very useful) salvage of cans was an essential prerequisite to the (extremely useful) salvage of railings, the recognition that collecting cans had occurred for show led many people to think that taking railings had been equally unnecessary. Some even suspected that the entire railings campaign had been a massive hoax.

Yet as Dawes had tried to point out in 1944, vast differences separated tins and railings. "You might as well compare margarine with butter. Railings are wrought or cast iron, and therefore superior raw material which is still in keen demand. Tin was an impurity which had to be dealt with when supplies were short."[30] Not everyone agreed with this disparaging view of tin cans, however. In response to Dawes's assertions, the director of the New London Electron Works pointed out that his firm, located in East Ham, consumed 600 tons of cans each week and turned

[28] *Parliamentary Debates*, Commons, 30 Nov. 1943, vol. 395, cols. 208–10.
[29] Graves, *Women in Green*, 206–7.
[30] "The Reason Why," *Public Cleansing and Salvage*, May 1944, 368.

them into steel plate for ships. "For the year ended 31st March 1943," he boasted, "this Company produced 23,000 tons of heavy steel and 165 tons of tin, all from old tins."[31] Yet few British firms – in sharp contrast to those in the United States – possessed the electrolytic technology necessary to strip the tin coating from steel cans.[32] Had they been able to do so, Britain's metal cans might have contributed materially to the war effort.

In August 1944, the Ministry of Supply added to the growing sense that recycling was no longer important by announcing that waste rubber could once again be treated as rubbish. Effective immediately, the Directorate of Salvage and Recovery would no longer pay local authorities for the rubber it had required them to collect.[33] This lack of advance warning struck many as unfair. They had devoted considerable time and expense to gathering rubber, but now they would receive no compensation for all the work they had done. The reversal of policy on household rubber collections, which followed the about-face on tin cans, suggested to some that the entire effort to salvage household items had been a pointless charade. Determined to quash such thoughts, the circular closed with the confusing assertion that "the new arrangements do not imply that there can be any relaxation in economy in the use of rubber and particularly tyres."[34]

In February 1945, J. C. Dawes, who had directed the salvage efforts of local authorities since the first months of the war, announced that the Ministry of Supply would soon relieve them of the obligation to collect any type of metal for salvage. At the same time, he noted that it remained "essential for waste paper, rags, household bones and kitchen waste to be collected in maximum quantities, not only during the remainder of the European war but for some time afterwards."[35] Such mixed messages, not surprisingly, prompted many members of the public to conclude that recycling was no longer important. After observing that there had

[31] Russell Webster, letter to the editor, *Public Cleansing and Salvage*, July 1944, 460.
[32] For a polemic against Britain's wartime steel industry, see Correlli Barnett, *The Audit of War: The Illusion and Reality of Britain as a Great Nation* (London: Macmillan, 1986), 87–106.
[33] Ministry of Supply, Salvage Circular 113 (Aug. 1944), Shakespeare Centre Library and Archive, Stratford-upon-Avon, BRR 55/14/31/4/4.
[34] *Parliamentary Debates*, Commons, 1 Aug. 1944, vol. 402, cols. 1204–5W.
[35] Ministry of Supply, Salvage Circular 118 (28 Feb. 1945), Walsall Local History Centre, 235/3.

"been a serious falling-off in the recovery of waste paper throughout the country," an official with the Directorate of Salvage and Recovery observed that "public apathy towards salvage continues to increase as the war situation improves, the general feeling apparently being that the materials are no longer required for war purposes."[36]

By the end of the war, the earlier idealism that many had felt for recycling had turned not to apathy but to disdain. An article published in *Public Cleansing and Salvage* in February 1945 stated, "For a time it was considered almost the highest form of domestic patriotism to separate salvage from un-economic waste. But since the announcement that salvaged food cans had no further value even the most conscientious of householders can hardly be expected to be enthusiastic any longer." Many people no longer associated salvage with conservation, but with wastefulness. Its prime symbol was the "unsightly and most unhygienic" village dump that had become, in the words of one observer, "an ignominious monument of a wasteful economy" and a threat to public health. Strewn with "unwanted tins, rags and paper," each dump served as "a perfect paradise for one of our worst disease carriers, the fly."[37]

PEACE AND PENURY

As the long war came to an end, the president of the Institute of Public Cleansing, A. L. Thomson, observed that "warfare is the supreme waste producer; it is destruction scientifically organised." This destruction had reached new heights during the war, which an anonymous contributor to *Public Cleansing and Salvage* referred to as "the most destructive orgy in history." To win it, Britain had expended enormous amounts of money and raw materials, to say nothing of lives. Even though the Axis threat had been eliminated, nearly six years of war had left the United Kingdom battered and nearly bankrupt. To recover from these losses, he suggested that Britain would have to "scrape and save" for a long time to come. "We have been consuming our mineral wealth on a prodigious scale. It is irreplaceable. What we have been doing is living off our capital. . . . It is

[36] Ministry of Supply, Directorate of Salvage and Recovery (L.A.), "Quarterly Bulletin for Salvage Officers in District No. 3," Feb. 1945, 1, Shakespeare Centre Library and Archive, Stratford-upon-Avon, BRR 55/14/31/4/4.

[37] "The Village Dump," *Public Cleansing and Salvage*, Feb. 1945, 206.

likely, therefore, that we shall be compelled to depend increasingly on renewal of natural materials and of usable wastes."[38]

Amid the celebrations to mark the defeat of Nazi Germany in May 1945, officials announced that the public would be allowed to have bonfires, but cautioned that they should not burn paper, for every available scrap was still needed for salvage.[39] Despite such admonitions, local authorities' paper collections had fallen to only a third of what they had been in 1943. Although it had little money to pay for imports, Britain was forced to resume its prewar practice of buying large quantities of wood pulp from Scandinavia.[40] Paper was in such short supply that the Ministry of Supply rationed the quantities it allowed to candidates during the 1945 general election campaign. Those with a constituency of 40,000 electors were allowed no more than one ton of paper for their campaigns, an amount that increased proportionally with every additional 10,000 voters. In addition, political parties were eligible to receive a ton of paper for each candidate they fielded. In the face of critiques that this tight rationing made it difficult for candidates to inform the public of their positions, Dawes insisted that "the matter is not a subject for debate."[41] Scientific journals also faced paper shortages, prompting Kenneth Sisam, head of Oxford University Press, to complain that the government was allowing such publications only 35 percent as much paper as they had used before the war.[42]

The Allied victory brought Britons no relief from austerity. In the short term, the state of its economy actually worsened. As soon as Japan surrendered, the United States cut off Lend-Lease aid to Britain, a decision that immediately plunged Britain back to the financially strapped circumstances that it had faced early in the war. Recycling offered a way for the

[38] A. L. Thomson, "What Is Waste? An Intriguing Question," *Public Cleansing and Salvage*, July 1945, 463; Diogenes, "Post-War Salvage – Yes or No?" *Public Cleansing and Salvage*, Feb. 1945, 205.

[39] "Litter and Salvage," *Justice of the Peace and Local Government Review*, 9 June 1945, 266.

[40] "Wood Pulp from Sweden," *Public Cleansing and Salvage*, July 1945, 463; Ministry of Supply, Directorate of Salvage and Recovery (L.A.), "Quarterly Bulletin for Salvage Officers in District No. 3," Dec. 1945, Shakespeare Centre Library and Archive, Stratford-upon-Avon, BRR 55/14/31/4/3.

[41] J. C. Dawes, "General Election (Paper Supplies)," June 1945, Shakespeare Centre Library and Archive, Stratford-upon-Avon, BRR 55/14/31/4/4.

[42] Kenneth Sisam to Frederick Lindemann, 27 June 1945, Cherwell Papers, Nuffield College Library, Oxford, C.75/3; Andrew Duncan to Frederick Lindemann, 20 July 1945, Cherwell Papers, Nuffield College Library, Oxford, C.78/4.

country to obtain essential raw materials without depleting its foreign currency reserves. Just as they had done during the war, officials argued that recycling was a civic responsibility – that its purpose was collective rather than individual.

British leaders believed that salvage would play an essential role in rebuilding their country both physically and economically. Said one official, "The war against waste has still to be waged with even greater vigour than before. The reconstruction of our national life depends mainly on the availability of raw materials, and the tendency to consign to the dustbin the very materials which are so urgently needed by many industries must be checked."[43] In late August 1945, less than two weeks after Japan surrendered, John Wilmot wrote to every local council in Britain to thank them for their contribution to the Allied victory. Wilmot, who had succeeded Andrew Duncan at the helm of the Ministry of Supply after the Labour Party won a landslide victory in the July 1945 general election, declared, "The large quantities of waste materials collected by Local Authorities have been of inestimable help in providing munitions of war and equipment, in lightening the strain on shipping and in meeting deficiencies in the supplies of important raw materials." After he praised the "zeal and perseverance" that people in most districts had devoted to salvage during the war, he cautioned that the need to recycle was "as great now as it has ever been." Wilmot asserted that "to re-establish our peace-time economy we still need the fullest possible collections of certain types of waste materials," including paper, rags, household bones, and kitchen scraps.[44]

This was most unwelcome news to many Britons, who by the end of the war were thoroughly tired of government exhortations for them to do their bit. During the war, thousands of women and children had acted as voluntary recycling workers, but local officials found it much more difficult to secure such unpaid assistance in the postwar period. Commenting on the sharp decline in recycling that had occurred, the West Midlands salvage official John Jorden asserted that "housewives – who, after all, form the backbone of the salvage effort – are very tired after nearly six years of war. They have had more than their share of worries

[43] John C. Jorden, "Salvage in the Midlands," n.d. [ca. 1946], 9, Shakespeare Centre Library and Archive, Stratford-upon-Avon, BRR 55/14/31/4/3. On Wilmot, see Ben Pimlott, "Wilmot, John Charles, Baron Wilmot of Selmeston (1893–1964)," in *Oxford Dictionary of National Biography* (Oxford: Oxford University Press, 2004).

[44] Wilmot's letter was sent to all local authorities under the cover of Ministry of Supply, Salvage Circular 121 (27 Aug. 1945), Walsall Local History Centre, 235/3.

and it is very natural for them to sit back and consider which of their tasks they can most easily set aside." Despite this trend, Jorden expressed the hope that most communities would continue to recycle their waste rather than send it to landfills or incinerators, "provided economic selling prices are assured."[45]

Jorden's caveat about the importance of making salvage financially attractive to local government was astute. The immediate postwar period brought tremendous fluctuations in prices for recyclable materials. During times of low demand, local authorities found it difficult, if not impossible, to find a buyer for the scrap that they collected. Under these circumstances, many local officials pressured central government to allow them to decide for themselves whether to operate recycling programs and what materials, if any, to collect. To encourage the continuation of paper recycling, in June 1945 the Directorate of Salvage and Recovery took the step of offering a substantial reward to every local authority that managed to collect at least 80 percent as much paper as it had during the corresponding period of the previous year.[46]

On 19 October 1945, Minister of Supply John Wilmot spoke about the salvage question on the BBC Home Service. After praising his listeners "for the great job you did during the war in collecting salvage," Wilmot called on them to persevere, "for the need today is just as great." Instead of turning waste into weapons, he announced that the time had come to "turn waste into wealth." The material that was needed most, he explained, was paper. "If every household would save an extra quarter of a pound of waste paper every week, it would make a tremendous difference to the output of our paper and board mills, which are now so very short of their raw material." Modifying the wartime practice of telling the public about specific munitions that could be made from recycled household articles, Wilmot told his listeners that "a Bradshaw Railway Guide, or a dozen penny newspapers, will provide enough material to make a couple of square feet of building board, or nine cartons to pack the cereals we have for breakfast." Wilmot warned listeners that Britain would face serious shortages for the foreseeable future. The crux of the problem was economic: "One dollar can buy only

[45] Ministry of Supply, Directorate of Salvage and Recovery (L.A.), "Quarterly Bulletin for Salvage Officers in District No. 3," June 1945, Shakespeare Centre Library and Archive, Stratford-upon-Avon, BRR 55/14/31/4/6.

[46] Ministry of Supply, Salvage Circular 120 (18 June 1945), Walsall Local History Centre, 235/3.

one dollar's worth, and if we waste our resources in bringing from abroad what we could save at home, we must go without something that otherwise we might have had."[47]

In a decision that heralded the postwar focus on economic health rather than military strength, in April 1946 the government transferred the Directorate of Salvage and Recovery from the Ministry of Supply to the Board of Trade.[48] Each ton of raw materials that Britain had to buy abroad added to its trade imbalance, and every ton that it was able to obtain at home helped to reduce that deficit. The public displayed little appetite for this view, however. As a contributor to the journal *Public Cleansing and Salvage* put it, most Britons were "anxious to cast off the shackles" of mandatory recycling and other regulations imposed in wartime. Although he shared the view that "controls are tiresome things to a free and democratic race," he maintained that it was too soon to abolish them. "The necessity for an effort equal to that made during the war is still with us," he insisted, "even though the urgency is not quite so obvious."[49]

In an attempt to chart a clear direction for government policy, the new Labour government appointed R. D. Fennelly, who had directed the wartime requisition of railings, to chair a committee on the status of recycling in Britain. Its report, issued in April 1947, painted a bleak picture. The amount of salvage generated by households had fallen sharply with the coming of peace, a decline that the committee attributed to the fact that individuals had become less "salvage conscious." In addition, many local officials felt that the requirement to collect recyclable materials had been "a wartime measure only. . . . They are in many cases, of course, handicapped in carrying out the double duty of refuse and salvage collection through their inability to obtain new vehicles; existing vehicles are old and constantly out of commission for repairs or tyre replacements."[50] One bright spot existed, however: free, or virtually free, scrap metal from Germany.

47 John Wilmot, transcript of radio address on the BBC Home Service, 19 Oct. 1945, Ulverston Urban District Council Records, Cumbria Records Office, Barrow, BSUD/U/S/Box 10.

48 Board of Trade, Salvage Circular 124 (10 Apr. 1946), Walsall Local History Centre, 235/3.

49 F. Shults, "The National Salvage Campaign," *Public Cleansing and Salvage*, May 1946: 459–60.

50 Interdepartmental Committee on Salvage, "Interim Report," 29 Apr. 1947, NA, MAF 35/725.

GERMAN SCRAP

To rebuild Britain's war-ravaged cities and industries, the nation needed steel. Imports of raw materials remained essential to the production of steel, as the United Kingdom possessed insufficient quantities of iron ore and scrap to supply its mills with all that they required. As Wilmot noted in 1947, "To meet the target of 14 million ingot tons of steel in 1948 . . . we must obtain over 1 million tons of scrap from external sources."[51] Unfortunately, however, Britain lacked the foreign currency needed to purchase scrap on the global market. Germany provided the answer. At the same time that the Soviets were stripping Eastern Germany of its industrial resources, the British were doing something similar – albeit on a smaller scale – in the parts of Germany that they occupied. Some of this metal was purchased, but most was seized as a spoil of war. "Booty Scrap," as an expert from the Ministry of Supply explained in 1948, referred to a wide range of materials: "scrap from enemy military stores, plant [i.e., equipment] from war factories and other warlike material, which is regarded as the property of the occupying Power."[52]

The British effort to acquire large quantities of booty scrap in Germany followed its frustration in persuading German merchants to sell commercial scrap at prices that the British considered reasonable (the official mentioned £6 a ton as a "fair price," but this was twice as much as the British had paid for the commercial scrap it had acquired in Germany from the end of the war through September 1946). Practically all of the scrap that Britain imported during the late 1940s came from Germany. As Wilmot explained to the prime minister, Clement Attlee, "Close and continuous contact has been maintained with the Foreign Office on this subject and the Foreign Secretary has impressed upon the Control Commission for Germany the paramount importance of securing these substantial scrap imports from Germany."[53] In a secret cable from Berlin in 1947, a British official informed London, "We are endeavouring to increase the supply of booty scrap. This is now at a rate of over 30,000 tons a month, and we are trying to work up to 50,000 to 60,000 tons. . . . It would not be wise to say too much about booty scrap. We believe that all we are proposing to take can appropriately be described as booty, but

[51] John Wilmot to Clement Attlee, 2 Oct. 1947, 5–6, NA, PREM 8/774.
[52] Scrap Investigation Committee, First Interim Report: Scrap from Germany, C.P. (48) 99, 6 Apr. 1948, 2, 3, NA, CAB 129/26.
[53] John Wilmot to Clement Attlee, 2 Oct. 1947, 5–6, NA, PREM 8/774.

we are not anxious to advertise the increased output for which we are working."[54]

Even when the British had to pay for German scrap, the victors skewed the terms heavily in their favor. Within the British occupation zone, the controlled price for ordinary scrap in 1948 was only 45 deutschmarks per ton ($15, or a little less than £4). As Sir Graham Cunningham, the former claims director of the War Damage Commission, observed, this price was too low to entice many German merchants "to part with a valuable capital asset which does not deteriorate and costs next to nothing to keep in stock."[55]

Adopting an extremely broad definition of "war booty," British occupation authorities seized thousands of tons of scrap each month before the creation of the Federal Republic of Germany in 1949. This practice led to considerable tension not only with the German government but also with the Truman administration, which by 1947 was more interested in strengthening West Germany as a bulwark against Soviet influence than in redressing Nazi crimes. Yet Britons, whose cities had been devastated by German bombs, considered it entirely reasonable to make Germans supply some of the resources needed for rebuilding their country.

A year after the war ended, the Salvage Branch of the British Army of the Rhine was busy sending large quantities of German munitions to Britain for melting and reuse. Although this material was supposed to be rendered harmless before leaving Germany, some live ammunition slipped through. In August 1946 H. W. Secker, the director of a company that funneled scrap from the British government to industry, informed the British Army's Salvage Branch in occupied Germany of

> two letters from Messrs. Colvilles Ltd. of Glasgow, who are receiving the greater part of the scrap shipped from Germany. They report two further explosions at their Hallside Works during the course of discharging battle-field scrap into their furnaces. In one case two men were badly injured, and in the other a furnace was completely wrecked, several men were injured, fortunately not fatally. Three days later there was a further explosion reported from their Dalzell Works and two explosions from their Lanarkshire Steel Works and a large piece of the roof of a steel furnace was blown out.[56]

[54] S. J. H., telegram to Foreign Office (German Section), 31 Oct. 1947, NA, T 172/2030.

[55] Scrap Investigation Committee, First Interim Report: Scrap from Germany, C.P. (48) 99, 6 Apr. 1948, 2, 3, NA, CAB 129/26.

[56] H. W. Secker to S. L. Hopkinson, 21 Aug. 1946, POWE 5/20.

Secker also sent a copy of his letter the Ministry of Supply. In a cover note, he suggested that something more than carelessness might be to blame. "We cannot overlook the possibility," he warned, "of German civilian labour on this job including explosive material as a deliberate attempt at sabotage."[57]

A secret government report, produced in 1948 under the direction of Sir Graham Cunningham, warned the cabinet that Britain might soon have to relinquish its claim on booty scrap: "There is . . . always the possibility that political objections may be raised to the dismantlement of factories and the destruction of plant which have been designated as booty. . . . Moreover, there are other allied European countries who might claim a share of this scrap. The Committee therefore urge that every effort should be made to speed up the removal and shipment of all available booty scrap while we have the opportunity."[58] Another source of difficulty arose in Britain's relations with the Soviet Union, which had already stripped its occupation zone of practically all industrial equipment and materials. In April 1948, at the start of their more than yearlong blockade of the western sectors of Berlin, the Soviets stopped the British from exporting scrap from the city.[59] After importing well over a million tons of free iron and steel in the form of booty scrap, as well as much smaller amounts of commercial scrap (much of which came from bomb-damaged buildings), shipments from Germany declined sharply after the founding of the Federal Republic in 1949.

On 6 September 1950, the minister of supply informed the cabinet that the Britain's steel output in the coming year would likely fall sharply because of the difficulty of obtaining scrap from abroad. As an official with the Ministry of Supply explained, the steady increase in British steel production since the end of the war had depended on scrap imports from Germany. Employing contorted language that reflected both a fear of economic collapse and a fear of making such concerns public, he urged the Ministry of Health to contact local authorities "for the purpose of emphasizing (without drawing attention in any way to the serious position with which we are faced – because that would not do at this point of time) the vital importance in the national interests of the Local Authorities

[57] H. W. Secker to T. Stanes, 22 Aug. 1946, POWE 5/20.

[58] Scrap Investigation Committee, First Interim Report: Scrap from Germany, C.P. (48) 99, 6 Apr. 1948, 2, 3, NA, CAB 129/26.

[59] "New Traffic Ban in Germany," *Times* (London), 21 Apr. 1948, 4.

taking every practicable step to see that all their surplus iron and steel material, and in particular scrap iron and steel, is collected and disposed of to the scrap merchants."[60]

In 1951, the West German government announced a "decree that 85% of the total scrap in the hands of the merchants must remain in Germany." British officials feared that the small amount to be exported would go primarily to the United States, due to German appreciation of American reconstruction assistance and a German preference to acquire dollars instead of pounds.[61] A short time later Humphrey Trevelyan, an official with the British contingent of the Allied Control Commission for Germany, informed the Foreign Office that "the Disposals Group are still collecting a considerable amount [of steel scrap] on Heligoland. We have, in fact, no real right to be doing so. Fortunately, neither the Federal Government nor the Land Government are challenging this." Britain was using the North Sea archipelago for bombing practice, and as long as it retained military control over Heligoland, the United Kingdom would be in a position to exploit its scrap. The letter ended with this revealing comment: "In view of the importance of scrap to the UK, we are trying to overcome every possible political objection in order that we can get the maximum amount from Germany."[62]

Six years after the fall of the Third Reich, one British official estimated that at least half a million tons of steel scrap remained to be cleared from bombed-out buildings in West Berlin alone. When the West German chancellor, Konrad Adenauer, visited London in December 1951, the minister of supply planned to offer British help in clearing this rubble, on the condition that Britain receive a large proportion of the steel that was recovered in the process.[63] This British approach toward West Germany contrasted sharply with the U.S. strategy. At the same time that the British were extracting resources from Germany in an effort to rebuild their own battered cities and struggling economy, the prosperous Americans, whose cities had been untouched by war, were showering the Federal Republic of Germany with enormous sums of Marshall Plan funds in an effort to make it a strong ally in the Cold War.

[60] W. M. Hill to H. F. Summers, 12 Oct. 1950, NA, HLG 102/94.
[61] "Notes of a Meeting Held on 22nd March, 1951 . . . ," NA, BE 2/172.
[62] Humphrey Trevelyan to Roger Stevens, 8 Oct. 1951, NA, FO 371/93512.
[63] "Note on the Subject of the Export of Steel Scrap from Western Berlin . . . ," 26 Nov. 1951, FO 371/93512; Ministry of Supply to Wahnerheide, 1 Dec. 1951 (draft), FO 371/93512.

RAILINGS

The London County Council lost over 7,000 tons of park railings during the war. In October 1946, it began negotiations with the Ministry of Supply, as did numerous other local authorities that had lost railings, in an attempt to negotiate compensation for them.[64] After nine months of discussion, the ministry offered the London County Council £80,000 as a final settlement of all its claims. Although this was far greater than the £9,000 the Council would have received according to the original compensation rate of 25 shillings per ton, it was only a fraction of the cost of replacing the railings that had been taken.[65]

Many churches also sought help in restoring their requisitioned railings. "Since the removal of the railings," complained a vicar in west London in 1946, "most of the trees + plants have been wilfully destroyed, the notice boards + church doors have been damaged + it is impossible to put anything in the garden without the risk of it being pulled up within a few hours."[66] Saint Anne's Church in Wandsworth, a borough in southwest London, faced similar problems. During the war the government had removed its railings and paid only £26 in compensation. Now, just a few years later, the church council found that replacing them would cost at least £1,000. It was essential to re-enclose the churchyard, explained the letter writer, because neighborhood children were playing and making loud noises right outside the church, which disrupted worship services.[67]

In September 1946 a group that represented several churches in the Liverpool area petitioned the Ministry of Works for assistance. "You may have recollections," it began, "that three or four years ago most of the iron railings were removed from the Churches in this Diocese of Liverpool; I don't know what use has been made of them, one report informs me that a great quantity are lying dumped on a vacant spot near St. Helens." The writer went on to explain that the removal of the railings had allowed "juvenile vandalism" to flourish and that churches had "spent literally thousands of pounds in trying to make good . . . the damage caused."[68]

64 G. L. Jordan to E. C. Wood, 16 Mar. 1947, NA, T 161/1405.

65 Cyril M. Walker, "Report by the Director of Housing and Valuer," 23 July 1947, London Metropolitan Archives, LCC/CL/CD/1/283.

66 Charles H. Drew to the archdeacon, 22 July 1946, NA, WORK 22/430.

67 E. A. Bellman to the minister of works, 26 July 1946, NA, WORK 22/430.

68 Churches Joint Advisory Committee for Liverpool, Bootle and Crosby to Ministry of Works, Manchester, 4 Sept. 1946, NA, WORK 22/430.

As a result of such problems, the government established a program to pay for replacement railings in cases where it considered them necessary. In the spring of 1947 it sent a memo to religious bodies throughout the country. Anxious to prevent a flood of claims, the Ministry of Works went to great lengths to keep the program secret. A cover letter asked recipients not to make any "reference to the scheme in diocesan and parochial publications." The memo also contained the following notice: "CONFIDENTIAL: THE CONTENTS OF THIS MEMORANDUM MAY NOT BE PUBLISHED OR COMMUNICATED TO THE PRESS WITHOUT THE PRIOR SANCTION OF THE DEPARTMENT." The reason for this secrecy was that the government had no intention of replacing all church railings. The Ministry of Works alone would decide which railings would be replaced with a new barrier, as well as the design and cost of the new fencing.[69] The secretive manner in which this program operated impeded a frank and transparent discussion of the problem and led to gross disparities. Although some churches received replacements under this program, most did not. In later years, individuals, churches, and government agencies spent vast sums replacing railings that the government had taken during the war. The cost of replacement was often a hundred times greater than what they had been offered in compensation.[70]

Similar secrecy and government sleight of hand prevented many individual property owners from obtaining compensation for railings requisitioned by the government. Because of their wartime responsibilities, millions of Britons were far from home when the government took their railings and thus learned of their removal long after it had occurred. The government eventually decided to accept claims submitted later than six months after requisition, but as one government official noted in an internal memorandum in November 1945, "No public announcement on the change of policy was ever made as it was considered unwise to give it any publicity."[71]

After public criticism continued to mount, officials retreated from their efforts to restrict new compensation claims. In April 1948, nearly three years after the war ended, the minister of works announced that anyone who had not previously accepted compensation for requisitioned railings

[69] Ministry of Works, "Memorandum on the Restoration of Gates and Railings Requisitioned from Churches, Chapels, and Other Buildings Used Primarily for Religious Services," Apr. 1947, NA, WORK 22/430.

[70] C. W. A. Millar, letter to the editor, *Daily Telegraph*, 16 Dec. 1967, 8.

[71] R. M. Hunter, memorandum, 5 Nov. 1945, NA, WORK 22/430.

or gates had until 1 June 1948 to file a claim. The rate it offered remained abysmally low: £1 per gate and 25 shillings per ton for cast iron railings (which it explained would amount to just 2 or 3 shillings for a typical house).[72]

The scars of war remained raw long after the war ended, as did opinions about requisition. In 1957, fifteen years after the government requisitioned the railings from Russell Square, Holborn Borough Council spent many thousands of pounds to erect new railings there.[73] As late as 1962, however, a chain link fence continued to mark the borders of Green Park, and chicken wire surrounded some other parks. In a strongly worded letter to the editor of the *Times* that year, architect and designer Misha Black argued that the time had come to replace the railings that had been taken during the war from London's parks and squares.[74] Not everyone agreed, however. Echoing arguments that some had made as early as the 1920s, one Londoner responded to Black's letter by asserting that railings represented an outmoded "civic and social concept" that had no place in an egalitarian Britain, adding that the railings were ugly. "It will be a breakthrough in the task of restoring the true charm of appearance to our capital when the last useless rail, fence, and barrier is torn up and carted off."[75]

THE END OF AN ERA

In October 1947, the Ministry of Supply urged councils to take a holistic view when they assessed the economics of salvage collection and waste disposal: "While a profit may not accrue from the sale of some materials a surplus may result from the disposal of others, such as waste paper for which comparatively high prices are obtainable." The circular mentioned a further factor, one that proponents of recycling continue to emphasize today: "Since salvage work . . . represents an expanded form of refuse collection and disposal, in assessing the cost of salvage operations, the alternative cost of collecting and disposing of the materials as normal refuse should be taken into account."[76]

[72] *Parliamentary Debates*, Commons, 14 Apr. 1948, vol. 449, cols. 76–7W; Ministry of Works, Regional Compensation Surveyor (Birmingham), Notes on compensation for the requisition of Gates and Railings [1948], Walsall Local History Centre, 235/5.

[73] "History Repeated in Russell Square," *Times*, 18 Apr. 1957, 12.

[74] Misha Black, letter to the editor, *Times* (London), 17 Nov. 1962, 9.

[75] J. G. Lomax, letter to the editor, *Times* (London), 28 Nov. 1962, 11.

[76] Board of Trade, Salvage Circular 130 (8 Oct. 1947), Ulverston Urban District Council Records, Cumbria Records Office, Barrow, BSUD/U/S/Box 1.

A clear continuity exists in the propaganda efforts that the British government employed to encourage recycling during and after the war. In both cases, posters, radio broadcasts, and films emphasized the national interest. Yet helping the nation achieve a more favorable balance of payments obviously lacked the drama, immediacy, and obvious importance of defeating Hitler. In January 1948, the Board of Trade asked the Central Office of Information (as the Ministry of Information became known after the war) to produce a one-reel film that would inform "the housewife" that the recycling of household paper helped to boost the country's exports.[77] A short time later Harold Wilson, president of the Board of Trade (and future Labour prime minister), called on local officials to increase their collections of paper "in view of the very serious economic position which confronts this country at the present time." Wilson also informed them that the Waste Paper Recovery Association would soon hold another national contest to promote paper recycling. "It is clear," he added, "that we shall need all the waste paper we can get for a long time to come and this is fully recognised by the paper and board mills in this country."[78]

Soon, however, Britain experienced a glut of waste paper, and many local authorities found it impossible to find buyers for the paper they had collected. In 1949 Harold Wilson abolished the regulations, in place since 1942, which made it illegal to destroy paper, rags, rope, or string. He also announced that cities and towns no longer had to operate recycling programs.[79] Finally, on 31 March 1950, the government formally disbanded the Directorate of Salvage and Recovery.[80]

Dawes viewed these changes with dismay. During its decade-long existence, the directorate (and its predecessor, the Salvage Department) had overseen the collection of more than eight million tons of paper, textiles, bones, scrap metal, and kitchen waste from the households

[77] M. L. G. Balfour to Director-General, 20 Jan. 1948, NA, INF 6/1245; T. A. O'Brien to John Shaw, 21 Jan. 1948, NA, INF 6/1245.

[78] Harold Wilson, circular letter to local authorities, 25 Feb. 1948, Walsall Local History Centre, 235/3.

[79] Harold Wilson, circular letter to local authorities, 27 Jan. 1949, Walsall Local History Centre, 235/3; "Waste Paper Supplies Have Overtaken Demand," *Walsall Observer*, 9 July 1949; Board of Trade, Salvage Circular 142 (29 June 1949), Walsall Local History Centre, 235/3.

[80] Board of Trade, Salvage Circular 144 (28 Feb. 1950), Walsall Local History Centre, 235/3. For an analysis of recycling during the subsequent quarter century, see Tim Cooper, "War on Waste? The Politics of Waste and Recycling in Post-War Britain, 1950–1975," *Capitalism Nature Socialism* 20, no. 4 (2009): 53–72.

TABLE 5. *Materials Salvaged by Local Authorities in England, Scotland, and Wales, Nov. 1939-Jan. 1950*

Kitchen Waste	3,261,704 tons
Waste Paper	2,863,673 tons
Scrap Metal	1,942,140 tons
Bones	74,361 tons
Rubber	30,438 tons[a]

[a] Rubber data are from Nov. 1939 through the end of 1946.
Sources: J. C. Dawes, "Report of the Salvage Operations of the Local Authorities in England/Wales and Scotland during the Period from 1st November, 1939 to 31st December, 1946," 3 Mar. 1947, MAF 35/725, "Directorate of Salvage and Recovery: Summary of 10 Years Work," n.d. [1950], NA, BT 258/405.

of Britain.[81] Having left the directorate a short time before Wilson ended compulsory recycling, the former salvage director did not hold back. In a speech to the Institute of Public Cleansing in February 1949, Dawes charged that Britain's "resources are being frittered away for want of imagination. . . . Unless something is done to bring home to all concerned the vital importance of this problem, we shall go on blindly for years in the same old groove, whilst in the process we shall be spending large sums of money and losing untold wealth in the form of raw materials."[82]

[81] "Directorate of Salvage and Recovery: Summary of 10 Years Work," n.d. [1950], NA, BT 258/405.
[82] J. C. Dawes, "Looking in the Mirror," 11 Feb. 1949, NA, HLG 51/825.

Conclusion

The war machine must be fed.

–*The Builder*, 1942[1]

The Second World War killed, maimed, and orphaned far more people than any other conflict in history. Compared to the incomprehensible suffering that this war brought about, it may seem perverse to focus on any other aspect of that war than the physical and mental anguish of its human victims. Yet we will never obtain a full realization of the wastefulness of war if we restrict our attention to its corporeal and psychic consequences. War destroys not only human bodies and minds but also cultural treasures, including buildings, books, manuscripts, historical artifacts, and works of art. It is tempting to blame all of the human suffering, property damage, and cultural destruction that comes with war on one's enemies, but whether we consider the treatment of dissenters and ethnic minorities, deaths caused by mistaking friend for foe, or the melting down of historical artifacts, a number of the wounds of the Second World War were self-inflicted.

In recent years, economists and policymakers have devoted increasing attention to the role of externalized costs in environmental problems, such as a decision by a factory owner to dump hazardous waste into a stream rather than treat it. In such a case, some of the costs of manufacturing are shifted to the ecosystem and the community – in the form of

[1] "Scrap Metal: Ministry of Works Drive," *Builder*, 16 Jan. 1942, 69.

contaminated water and harm to fish, for instance. When this happens, the expense required to clean up a problem often exceeds by many times the cost of measures that would have prevented it in the first place. Government policies can reduce the externalization of costs by requiring those who create wastes to pay for their proper disposal. Recycling is often the most environmentally sustainable way to deal with waste materials because it reduces pollution to the air, land, and water; conserves limited resources; and saves energy.

In wartime, however, people have long pursued recycling for quite different reasons. Few political or military leaders – past or present – have devoted much thought to the long-term environmental consequences of their actions, but they *have* recognized that in wartime they could benefit, at least in the short run, by modifying the environment to their advantage, causing harm to their enemies, and conserving the material and economic resources at their disposal. Although recycling has played a part in warfare since well before the industrial revolution, it was not until the First World War that warring states began to organize systematic programs to transform the waste produced by civilian consumption into weapons of war. These efforts grew in importance during the Second World War, when all major belligerents made extensive use of recycled materials.

Between 1939 and 1945, recycling held a particularly important place in Britain, where officials employed it in support of goals that were simultaneously strategic, diplomatic, psychological, and economic. In wartime Britain, recycling functioned not as a way to *prevent* the externalization of environmental harm but rather as a means to *promote* the externalization of the costs of military production. The government did this through a combination of requisition, price controls, and volunteer labor, all of which shifted some of the costs associated with acquiring raw materials to the people who provided them.

The Second World War was incredibly wasteful and expensive, yet the enormous stakes involved led most people to disregard ordinary considerations: if something provided a military advantage, most believed that it had to be pursued, no matter how high the cost in money or in infringements of personal liberty. Toward this end, the British government took a wide range of steps to intervene in the market. These measures included rationing, price controls, import and export restrictions, licensing of merchants, and subsidies to local government. Recycling mattered for reasons that transcended cost-benefit analysis: as an investment in national survival and victory, the government expected everyone to recycle even if salvage failed to generate a short-term profit.

Wartime recycling differed fundamentally from peacetime recycling in another aspect as well. Under normal circumstances, recycling contributes to environmental sustainability because it reduces waste, but the recycling that took place in Britain during the Second World War was unsustainable on at least three levels: it consumed items of irreplaceable cultural, historical, and personal significance; it resulted in the permanent loss of metals and other resources when the bombs and shells made from them exploded; and the use of these weapons destroyed lives, property, and cultural artifacts.

Scrap drives absorbed not only resources that were part of the normal waste stream but also objects of value that few persons would have considered discarding in peacetime. Despite the often-stated goal of preventing waste and making useless items into war materiel, Britain's wartime salvage campaign destroyed a significant portion of the country's historical inheritance. Resistance to the recycling of railings, books, and personal papers often arose when their owners refused to accept the government's assertion that these items should be treated as scrap.

Wartime officials often acted as if recyclable materials were free resources, but this obscured the true costs of acquiring them. Salvage depended on armies of unpaid volunteers and smaller numbers of paid workers, most of whom were employed by local government, to gather, sort, clean, bundle, and transport the materials that people contributed. To extract this labor in wartime, the government found it necessary to use a combination of moral pressure, financial incentives, and coercion.

The story of wartime salvage provides an important view of the role of children in war not simply as victims but also as agents who helped to arm their nation's warriors. Looking back at wartime recycling from the perspective of 1949, one local official noted that "children often made a nuisance going round people's yards and got hold of stuff which sometimes could not be classed as salvage. There was Police criticism; school playgrounds got littered and altogether the energies of the children were misdirected."[2] This assessment ignored the many significant contributions that children made to wartime salvage efforts and reflected a bias in favor of municipal as opposed to voluntary efforts. Dawes, however,

[2] Walsall Public Works Department, "Salvage Collections in Schools," 31 Jan. 1949, Walsall Local History Centre, 235/3.

credited children with bringing in 80 percent of the books salvaged in Britain during the war.[3]

From the beginning of the Second World War, most Britons accepted the notion that recycling was a patriotic duty. Never before had the British people recycled with so much zeal or thoroughness, and no comparable recycling effort has occurred in Britain since then. This involved a major paradigm shift, for prior to the Second World War, virtually everyone considered the conservation or reuse of materials entirely from the perspective of personal gain or loss. The extraordinary nature of wartime recycling struck most participants as a unique case; peacetime, they assumed, would bring an end to recycling.

Another postwar change, one that happened more slowly than the psychological shift from a wartime mentality to a peacetime mindset, was a growing feeling that economic planning had gone too far and that laissez-faire principles should be at least partially restored. Even as it nationalized the coal, steel, gas, and transport industries, the postwar Labour government abolished subsidies to local authorities that encouraged recycling. Finally, in 1954, the Conservatives, once again in power, ended rationing of consumer goods. To many, this moment symbolized the end of a quarter century of frugality and sacrifice that had begun with the Great Depression and continued throughout the war and its aftermath. As the day approached when rationing would end, Lord Woolton, who had overseen the rationing of food and other necessities during the war, announced that he looked forward to throwing his own ration book into the fire in celebration. Woolton's remarks prompted an indignant reprimand from another official, who declared that even though the nation no longer had to rely upon rationing, the United Kingdom could not afford to destroy paper.[4] Few Britons followed this admonition, for they thought of recycling the same way they viewed make-do and mend, meatless meals, and of course rationing: as an unpleasant yet necessary wartime measure that had no place in a free and prosperous society.

For the next quarter century, recycling receded into the background for most Britons. People born after the war grew up in what came to be

[3] J. C. Dawes, memorandum, 21 June 1944, Walsall Local History Centre, 235/3.

[4] Woolton deflected the criticism by insisting that he had been misquoted. Instead of destroying his ration book, he announced that he intended to keep it as a memento. See J. K. T. Frost to Sidney T. Garland, 8 July 1954, NA, BT 64/5087; Michael D. Kandiah, "Marquis, Frederick James, First Earl of Woolton (1883–1964)," in *Oxford Dictionary of National Biography* (Oxford: Oxford University Press, 2004).

called the throw-away society. Attitudes toward waste and leftovers proved an important generational marker. Paradoxically, it would be the postwar baby boomers who would later embrace recycling for reasons very different from those that their parents and grandparents had embraced during the war years. Recycling, when it reemerged in the public consciousness in Britain around the time of the first Earth Day in 1970, was not an act of wartime sacrifice or national reconstruction. Quite to the contrary, during the early 1970s many people in Britain considered recycling something of a countercultural phenomenon. Many of those who embraced recycling were part of the student movement, the women's movement, the anti-war movement, and the environmental movement.[5]

Commenting on the decline in local waste paper collection during the final years of the war, historian Tim Cooper notes that "there was general pessimism about the ability to maintain the level of collections after the war."[6] Implying that everything recycled during the war would have belonged to the waste stream under normal circumstances, Cooper treats wartime recycling statistics as representing an ideal that could and should have been maintained after the war. Geographer Mark Riley similarly posits that the rise and fall of recycling during the Second World War was a consequence of public enthusiasm during the early years of the conflict, followed by "waning interest . . . as the war progressed and almost complete cessation as the war ended."[7] Diminishing public enthusiasm no doubt contributed to the decline in recycling that took place during and after the war, but it would be simplistic to read this decline as evidence of a return to prodigality. Material factors also influenced behavior. By the middle of the war, Britain had experienced several years of restricted consumption, coupled with assiduous salvage efforts. As a result, by 1943, relatively few items remained in Britain that could be exploited as sources of scrap.

Another factor that must be considered is the volatility typical of reclaimed materials markets in the absence of stabilizing mechanisms such as government price controls. Consumers of large quantities of waste paper, such as the publishing and packaging industries, obviously have an interest in keeping the price of paper as low as possible. Yet the waste paper merchants who supply them face a dilemma. On the one

[5] J. R. McNeill, *Something New under the Sun: An Environmental History of the Twentieth-Century World* (New York: W. W. Norton, 2000), 336–40.
[6] Cooper, "Challenging," 730. [7] Riley, "From Salvage to Recycling," 86.

hand, they need to have enough paper to meet their customers' demands; on the other, they do not want stocks of waste paper to become so plentiful as to drive down the price they can charge. Their profits, in other words, depend on an equilibrium between supply and demand. If paper is too scarce they will not have enough to sell, but if it is too plentiful they will not be able to earn a profit, and those who collect will not be able to cover the costs of doing so. As the economists Frank Ackerman and Kevin Gallagher have noted, similar price swings in the waste paper market have occurred in recent decades, with detrimental effects on the success of recycling programs. Their analysis led them to conclude that government regulation of prices, rather than interfering with the efficient operation of a supposedly wise "invisible hand" of the free market, actually makes it work better by exerting a moderating influence on the herd mentality among commodities investors.[8]

People often idealize the Second World War era as a time in Britain of class harmony, equality of sacrifice, and an absence of violent crime – a phenomenon that the historian Angus Calder described as "the myth of the Blitz." Environmentalists have created their own myths about wartime Britain as experiencing a golden age of recycling, thus ignoring the cultural, economic, social, and even environmental costs that recycling entailed in a time of total war. In peacetime, the monetary value of copper and steel means that few large articles containing these metals are discarded. In wartime, however, vast quantities of these materials are blown up or sent to the bottom of the ocean. As this book has shown, the salvage campaigns of the Second World War made significant contributions to the Allied victory, but, like the war itself, they trampled on individual rights and destroyed much that was of value. This may have been unavoidable under the circumstances, but it was far from an ideal that ought to be emulated, and it raises important questions about the proper boundaries between government power and civil liberties, particularly in times of crisis.

As the Second World War neared its bloody conclusion, some of those involved most closely in Britain's wartime salvage efforts began to reflect on the broader significance of their work as well as the paradoxical linkages between conservation and destruction during total war. One of the most thoughtful commentators on this subject was the writer

[8] Frank Ackerman and Kevin Gallagher, "Mixed Signals: Market Incentives, Recycling, and the Price Spike of 1995," *Resources, Conservation & Recycling* 35 (June 2002): 275–95.

Sir James Marchant, who from 1941 to 1944 worked for the Directorate of Salvage and Recovery in its efforts to collect railings.[9] Speaking at the Oxford Union in 1943, Marchant contrasted the ideals of the Atlantic Charter to the existing inefficient use and inequitable distribution of natural resources, which he saw as a major source of world conflict. He argued that the prevention of waste would not only help to decide the outcome of the present war but would play an essential role in postwar reconstruction and efforts to build a lasting peace. The efficient use of raw materials, he argued, provided "an opportunity – it may be our last – to overthrow the God of Waste in home and factory, in work and leisure, in health and education, in State and Church. We shall help to preserve the peace of the world, for which many are suffering and dying, by ceasing to waste."[10]

Most people in Britain initially rejected this advice. Although the Second World War changed the way people thought about recycling, most viewed wartime salvage as a temporary emergency measure akin to rationing. So strong was the association between recycling and wartime sacrifice in the minds of both political leaders and ordinary citizens that the nation soon returned to prewar patterns of waste disposal. Several decades would pass before those who had been too young to fight in the war transformed the notion of recycling from a military pursuit intended to benefit one's own nation to an activity intended to promote environmental sustainability on a global level.

[9] Born in 1867, Marchant was already in his seventies when the war began. He died in 1956 at the age of 88. For his obituary, see "Sir J. Marchant," *Times* (London), 22 May 1956, 11.

[10] James Marchant, *World Waste and the Atlantic Charter* (Oxford: Blackwell, 1943), 4, 14, 16.

Bibliography

Archives

Bodleian Library, Oxford University, Department of Special Collections and Western Manuscripts

Papers of Christopher Addison
Papers of Harold Macmillan
Papers of Frederick James Marquis, first Earl of Woolton

Borthwick Institute for Archives, University of York

Papers of Edward Frederick Lindley Wood, first Earl of Halifax

British Library, London

British Library Corporate Archives

Churchill Archives Centre, Churchill College, Cambridge

Winston Churchill Papers

Cumbria Records Office, Barrow

Ulverston Urban District Council Records

East Sussex Record Office, Brighton

Archive of Charles Sheppard and Sons of Battle, Solicitors
East Sussex County Council Records

Franklin D. Roosevelt Library (National Archives and Records Administration), Hyde Park, New York

President's Official File
Papers of James H. Rowe

Georgetown University Library Special Collections Research Center, Georgetown University, Washington, D.C.

Harry L. Hopkins Papers

Library of Congress Manuscript Division, Washington, D.C.

W. Averell Harriman Papers

London Metropolitan Archives

London County Council Records
Middlesex County Council Records

Modern Records Centre, Warwick University, Coventry

Trades Union Congress Records

U.S. National Archives and Records Administration (NARA), College Park, Maryland

Office of Lend-Lease Administration Records

National Archives of England, Wales, and the United Kingdom, Kew (NA)

Board of Education: ED 10, 11, 138
Board of Trade: BT 11, 28, 64, 87, 213, 258
Cabinet Office: CAB 68, 75, 115
Exchequer and Audit Department: AO 30
Foreign Office: FO 371, 837, 954
General Register Office: RG 23
Home Office: HO 186, 207
Metropolitan Police: MEPO 2
Ministry of Agriculture and Fisheries: MAF 35, 58
Ministry of Aviation: AVIA 11, 15, 22, 38, 46
Ministry of Housing and Local Government: HLG 51, 102
Ministry of Information: INF 1, 6
Ministry of Munitions: MUN 4
Ministry of Power: POWE 5
Ministry of Supply: SUPP 3, 14

Ministry of War Transport: MT 59
Ministry of Works: WORK 14, 16, 19, 22
Prime Minister's Office: PREM 1, 3, 8
Public Record Office: PRO 30
Treasury: T 160, 161, 162, 172
War Office: WO 107

National Library of Wales, Aberystwyth

Papers of Megan Lloyd George

Nuffield College Library, Oxford University

Papers of Frederick Alexander Lindemann, Viscount Cherwell of Oxford

Parliamentary Archives, London

Papers of William Maxwell Aitken, first Baron Beaverbrook

Royal Voluntary Service Archive & Heritage Collection, Devizes

Records of Women's Voluntary Services

Shakespeare Centre Library and Archive, Stratford-upon-Avon

Reformed Borough of Stratford-upon-Avon Records

University College London, Special Collections

Library Association Records

University of Virginia Library, Special Collections, Charlottesville, Virginia

Edward R. Stettinius Jr. Papers (Accession #2723-z)

Walsall Local History Centre

Files of Correspondence concerning Salvage during World War II (235/1–5)

The Women's Library, London School of Economics

Records of the National Federation of Women's Institutes

Oral History

Jackson, Evelyn. Interview by Peter Thorsheim. St. Leonards-on-Sea, 27 July 2009.

Newspapers

Bath Weekly Chronicle and Herald
Daily Telegraph
Devon and Exeter Gazette
Farmer and Stockbreeder
Gloucestershire Echo
Graphic (London)
Hastings & St. Leonards Observer
Manchester Guardian
New York Times
News Chronicle
Press and Journal (Aberdeen)
Spectator
Sunday Express
Sunday Times (London)
Times (London)
Tottenham and Edmonton Weekly Herald
Tottenham Calling
Tribune (London)
Walsall Observer
Walsall Times
A Week of the War
Western Daily Press and Bristol Mirror
Western Morning News (Devon)
Yorkshire Post

Contemporary Publications

"Aircraft from Pots and Pans." *Waste Trade World and the Iron and Steel Scrap Review*, 20 July 1940, 2–3.
The Amazing Story of Weapons Made from Waste. London: Odhams, 1942.
"Ancient Records for Pulp?" *Waste Trade World and the Iron and Steel Scrap Review*, 23 Nov. 1940, 15.
"The Appeal for Aluminium." WVS *Bulletin*, Aug. 1940, 1.
"Arms and the Scrap." *Waste Trade World and the Iron and Steel Scrap Review*, 22 Feb. 1941, 2.
"A Big Task: Where the Cleansing Service Comes in." *Public Cleansing*, Jan. 1940, 144.
"Bones for War Purposes." *Waste Trade World and the Iron and Steel Scrap Review*, 17 Aug. 1940, 3.
"Book Recovery." *Library Association Record* 45 (Oct. 1943): 174.
"Books Wanted." *Public Cleansing and Salvage*, Feb. 1943, 195.
"Britain Buys U.S.A. Scrap." *Waste Trade World and the Iron and Steel Scrap Review*, 31 Aug. 1940, 3.
Central Parish Councils Committee. *Salvage in Rural Areas*. London: National Council of Social Service, 1944.

Chudleigh, C. A. E. "Apathy and Waste." *Spectator* 164 (23 Feb. 1940): 245.
City of Bath Salvage Scheme. Bath: n.p., n.d. [1941].
"Control of War Materials." *Waste Trade World and the Iron and Steel Scrap Review*, 9 Sept. 1939, 2.
Crowther, Geoffrey. *Ways and Means of War*. Oxford: Clarendon Press, 1940.
"Crystal Palace North Tower Felled." *Waste Trade World and the Iron and Steel Scrap Review*, 19 April 1941, 9.
Dawes, J. C. "Making Use of Waste Products." *Journal of the Royal Society of Arts* 90 (15 May 1942): 388–408.
"Notes on the Work of the Ministry of Supply Salvage Department and Municipal Salvage." *Public Cleansing*, Sept. 1941, 8–14.
Report of an Investigation into the Public Cleansing Service in the Administrative County of London. London: HMSO, 1929.
deWilde, John C., James Frederick Green, and Howard J. Trueblood. "Europe's Economic War Potential." *Foreign Policy Reports* 15 (15 Oct. 1939): 178–92.
Diogenes. "Post-War Salvage – Yes or No?" *Public Cleansing and Salvage*, Feb. 1945, 205–6.
"The Distribution of Raw Materials." *Britain To-Day*, 31 March 1939, 6–7.
"Drive for Scrap Metal: Stocks of Three Tons or More Must Be Disclosed." *Public Cleansing and Salvage*, Oct. 1942, 64.
Earley, James S., and William S. B. Lacy. "British Wartime Control of Prices." *Law and Contemporary Problems* 9, no. 1 (1942): 160–72.
"Edinburgh's Book Salvage Drive Begins." *Public Cleansing and Salvage*, Feb. 1943, 216.
"Edinburgh's Campaign." *Public Cleansing*, April 1940, 259.
"Editorial." *Public Cleansing*, April 1942, 208.
"Editorial." *Public Cleansing and Salvage*, Oct. 1942, 44.
"End of Crystal Palace North Tower." *Waste Trade World and the Iron and Steel Scrap Review*, 26 April 1941, 4–5.
"End of Palace Tower." *Waste Trade World and the Iron and Steel Scrap Review*, 29 March 1941, 2–3.
"'Find' in Waste Paper." *Waste Trade World and the Iron and Steel Scrap Review*, 7 Dec. 1940, 9.
Flower, C. T. "Manuscripts and the War." *Transactions of the Royal Historical Society*, 4th ser., 25 (1943): 15–33.
Forbes, James, ed. *The Municipal Year Book and Encyclopædia of Local Government Administration, 1944*. London: Municipal Journal, 1944.
Forsdyke, John. "The Museum in War-Time." *British Museum Quarterly* 15 (1941–1950): 1–9.
Galway, J. Letter to the editor. *Waste Trade World and the Iron and Steel Scrap Review*, 17 Aug. 1940, 8.
"Germany Wastes Nothing." *Public Cleansing*, Dec. 1937, 146.
"Get Those Skins." *Waste Trade World and the Iron and Steel Scrap Review*, 8 Nov. 1941, 2.
"Getting a Move On." *Public Cleansing*, Jan. 1940, 120.
"Government and Essential Supplies." *Waste Trade World and the Iron and Steel Scrap Review*, 26 Aug. 1939, 2.

"Government Salvage." *Waste Trade World and the Iron and Steel Scrap Review*, 25 Nov. 1939, 3.
Great Britain. Ministry of Economic Warfare. *The Bomber's Baedeker (Guide to the Economic Importance of German Towns and Cities), Part III: Survey of Economic Keypoints in German Towns and Cities (Population 15,000 and Over)*. [London]: Ministry of Economic Warfare, 1943.
Great Britain. Ministry of Home Security. *Report on an Inquiry into the Accident at the Bethnal Green Tube Station Shelter on the 3rd March, 1943*. London: HMSO, 1945.
Great Britain. Ministry of Information. *Wartime Social Survey: Salvage*. [London]: n.p., 1942.
Great Britain. Ministry of Supply. *Memorandum on Salvage and Recovery*. London: HMSO, 1944.
Great Britain. Ministry of Supply. *Salvage: Lectures to Schools and Test Papers (With Answers)*. [London]: Ministry of Supply, 1942.
Great Britain. Ministry of Supply. *The School Salvage Steward's Guide*. [London]: Ministry of Supply, 1942.
Great Britain. Parliament. *Parliamentary Debates (Hansard)*. London: HMSO, various years.
Great Britain. *Statutory Rules and Orders 1940*, vol. 2. London: HMSO, 1941.
Great Britain. *Statutory Rules and Orders 1942*, vol. 2. London: HMSO, 1943.
"A Great Role—Stopping the Leaks in National Waste." *Public Cleansing*, May 1940, 267–8.
Gurney, H. Letter to the editor. *Waste Trade World and the Iron and Steel Scrap Review*, 12 Oct. 1940, 2.
"Hastings Prosecution." *Public Cleansing and Salvage*, Nov. 1943, 128.
Hill, Reginald Harrison. "The National Central Library: Impressions and Prospects." *Library Association Record* 47 (Dec. 1945): 246–52.
"The House of Commons." *Builder*, 6 Feb. 1942, 120.
"The Institute of Public Cleansing: Scottish Centre at Glasgow." *Public Cleansing*, June 1940, 313–16.
Institute of Scrap Iron and Steel. *Yearbook*. Washington, D.C., 1939, 1940, 1942.
"Iron and Steel Scrap Markets." *Waste Trade World and the Iron and Steel Scrap Review*, 25 March 1939, 13.
"Iron Railings." *Waste Trade World and the Iron and Steel Scrap Review*, 31 Aug. 1940, 19.
"Islington in the Forefront." *Public Cleansing*, June 1940, 335.
Jenkinson, Hilary. "British Archives and the War." *American Archivist* 7, no. 1 (1944): 1–17.
"The King's Scrap Gift." *Waste Trade World and the Iron and Steel Scrap Review*, 20 July 1940, 5.
Leach, Benny. Letter to the editor. *Waste Trade World and the Iron and Steel Scrap Review*, 17 Aug. 1940, 8.
Leadlay, E. O. "Waste Paper Goes to War." *Waste Trade World and the Iron and Steel Scrap Review*, 18 July 1942, 22.
"London Scrap Exhibition." *Waste Trade World and the Iron and Steel Scrap Review*, 17 Aug. 1940, 13.

"Making Manchester Salvage Conscious." *Waste Trade World and the Iron and Steel Scrap Review*, 22 Nov. 1941, 10–11.
"Man Who Cannot Read Is Now Salvage Sleuth." *Public Cleansing*, Jan. 1942, 146.
Marchant, James. *World Waste and the Atlantic Charter*. Oxford: Blackwell, 1943.
Marinetti, F. T. "The Founding and Manifesto of Futurism" (1909). In *Marinetti: Selected Writings*, edited by R. W. Flint, translated by R. W. Flint and Arthur A. Coppotelli, 39–44 London: Secker & Warburg, 1972.
"Marlborough Guns." *Waste Trade World and the Iron and Steel Scrap Review*, 7 Sept 1940, 2.
Mayhew, Henry. *London Labour and the London Poor*. 4 vols. London: Charles Griffin, 1851–61.
"Memorial Plaque for Salvage." *Public Cleansing*, April 1942, 224.
"Miss Megan Lloyd George, M.P., Cuts Railings for Scrap Campaign." *Waste Trade World and the Iron and Steel Scrap Review*, 28 Dec. 1940, 8.
Moss, C. R. "Reclamation of Waste Materials from Refuse in War Time." *Public Cleansing*, Jan. 1940, 132–8.
"Mr. Dawes at Northampton." *Public Cleansing*, Feb. 1941, 172.
Murray, David. "Railings." *Public Cleansing*, Jan. 1941, 162–3.
National Central Library. *26th and 27th Annual Report of the Executive Committee, 1941–42 and 1942–43*. London: n.p., 1943.
"The National Central Library Fire." *Library Association Record* 43 (May 1941): 88.
"National Scrap Campaign" *Waste Trade World and the Iron and Steel Scrap Review*, 23 Sept. 1939, 2.
"National Survey of Fixed and Demolition Scrap." *Builder*, 11 Oct. 1940, 367.
"The National Survey of Fixed and Demolition Scrap Iron and Steel: Results of First Month's Test Operation." *Public Cleansing*, Dec. 1940, 118.
"Naval Relics Saved." *Waste Trade World and the Iron and Steel Scrap Review*, 24 Aug. 1940, 2.
"Nazi Vandals." *Waste Trade World and the Iron and Steel Scrap Review*, 23 Aug. 1941, 3.
"New Controller of Salvage (Local Authorities)." *Public Cleansing*, Jan. 1942, 126.
"Old Guns for the Melting Pot." *Waste Trade World and the Iron and Steel Scrap Review*, 10 Aug. 1940, 8.
O'Neil, B. H. St. J. "Historic Buildings and Enemy Action in England." *Journal of the Royal Institute of British Architects*, March 1945, 132–3.
"Park Railings: Manchester's Report on Removal." *Municipal Review*, Sept. 1940, 190.
Pickford, H. Letter to the editor, *Waste Trade World and the Iron and Steel Scrap Review*, 24 Aug. 1940, 7.
Plummer, Alfred. *Raw Materials or War Materials?* London: Victor Gollancz, 1937.
"Precautionary Measure." *Waste Trade World and the Iron and Steel Scrap Review*, 20 July 1940: 3.

"Preservation of Iron Railings of Architectural and Historic Interest." *Journal of the Royal Institute of British Architects* 49 (Nov. 1941): 16.
"Preservation of Iron Railings of Interest." *Builder*, 14 Nov. 1941, 434.
"Prevention of Waste in Germany." *Public Cleansing*, April 1940, 259.
"Public Salvage a Vital Necessity." *Public Cleansing and Salvage*, Sept. 1943, 54.
"Questions in the House." *Public Cleansing*, March 1941, 224.
Rag Man. "Outshots." *Waste Trade World and the Iron and Steel Scrap Review*, 14 Jan. 1939, 1.
"Outshots." *Waste Trade World and the Iron and Steel Scrap Review*, 16 Sept. 1939.
"Outshots." *Waste Trade World and the Iron and Steel Scrap Review*, 25 Nov. 1939, 2.
"Outshots." *Waste Trade World and the Iron and Steel Scrap Review*, 17 Aug. 1940, 1.
"Railings for Scrap." *Builder*, 17 May 1940, 581.
"Railings for Scrap: New Campaign." *Builder*, 8 Nov. 1940, 463.
"Railings for War Weapons." *Builder*, 17 Oct. 1941, 344.
"Railings Protest." *Public Cleansing*, April 1942, 224.
"Railings Removal." *Public Cleansing and Salvage*, Dec. 1942, 123.
"The Reason Why." *Public Cleansing and Salvage*, May 1944, 368.
"Removal of Railings." *Municipal Review*, Feb. 1942, 23.
"Requisitioned Railings." *Public Cleansing and Salvage*, Aug. 1943, 424.
Robertson, D. J. W. "Is Pre-Separation of Salvage Really Practicable?" *Public Cleansing*, April 1941, 244.
Roush, G. A. *Strategic Mineral Supplies*. New York: McGraw-Hill, 1939.
"Salvage." WVS *Bulletin*, March 1940, 8–9.
"Salvage." WVS *Bulletin*, July 1940, 4.
"Salvage." WVS *Bulletin*, Nov. 1940, 4.
"Salvage and the Utilization of Waste." *Nature*, 29 June 1940, 988–9.
"Salvage Department." *Waste Trade World and the Iron and Steel Scrap Review*, 23 Dec. 1939, 4.
"A Salvage Exhibition." *Public Cleansing*, April 1940, 257.
"Salvage Facts." WVS *Bulletin*, Dec. 1942, 5.
"Salvage News." WVS *Bulletin*, Feb. 1943, 2.
"Salvage of Paper." *Municipal Review* 13 (March 1942): 48.
"Salvage of Refuse." *Municipal Review*, Dec. 1939, 394.
"Salvage of Waste." *Municipal Review*, May 1940, 108.
"Salvage Work: No 'Glamour' But . . . " *Public Cleansing and Salvage*, Aug. 1944, 494.
"Salvaging Waste Will Be Compulsory in Ulster." *Public Cleansing*, Jan. 1941, 156.
Sayers, W. C. Berwick. "Britain's Libraries and the War." *Library Quarterly* 14, no. 2 (1944): 95–99.
"Schools Salvage Campaign." *Municipal Review*, June 1940, 118.
"Scottish Centre Annual Meeting." *Public Cleansing*, May 1942, 238–44.
"Scrap for Britain." *Waste Trade World and the Iron and Steel Scrap Review*, 23 Dec. 1939, 2.

"Scrap for Victory." *Waste Trade World and the Iron and Steel Scrap Review*, 13 July 1940, 2.
"Scrap from the Sky." *Waste Trade World and the Iron and Steel Scrap Review*, 3 Aug. 1940, 4.
"Scrap Iron Helps War – and Rates." *Public Cleansing*, Dec. 1940, 131.
"Scrap Metal Shortage." *Public Cleansing and Salvage*, Feb. 1943, 212.
"The Scrap Position in Britain and Germany." *Waste Trade World and the Iron and Steel Scrap Review*, 16 Dec. 1939, 10.
"Scrap Shortage." *Waste Trade World and the Iron and Steel Scrap Review*, 8 April 1939, 7.
"Scrap the Railings." *Waste Trade World and the Iron and Steel Scrap Review*, 9 Nov. 1940, 2.
"Shorter Notes." *Economist*, 30 Aug. 1941, 262.
"Should the Trade Close Down?" *Waste Trade World and the Iron and Steel Scrap Review*, 20 July 1940, 2.
Shults, F. "The National Salvage Campaign." *Public Cleansing and Salvage*, May 1946: 459–60.
"Sidelights on the Scrap Front." *Waste Trade World and the Iron and Steel Scrap Review*, 31 Aug. 1940, 22.
Silverman, Jack "Why Not Fix Minimum Prices?" *Waste Trade World and the Iron and Steel Scrap Review*, 24 Aug. 1940, 5–6.
"Give the Tatter a Square Deal." *Waste Trade World and the Iron and Steel Scrap Review*, 21 Sept. 1940, 13–14.
"Maximum Co-Operation Is Essential." *Waste Trade World and the Iron and Steel Scrap Review*, 16 Nov. 1940, 5–6.
"Co-Operate in the National Interest: Welcome, Mr. Bell!" *Waste Trade World and the Iron and Steel Scrap Review*, 30 Nov. 1940, 6–8.
"Son's 500 Letters as Salvage." *Public Cleansing*, March 1942, 194.
"Statistic Department." WVS *Bulletin*, Nov. 1939, 7.
"Statistical Survey of W.V.S. Enrolments." May 1940, 2.
Stettinius, Edward R. Jr. *Lend-Lease: Weapon for Victory*. New York: Macmillan, 1944.
"Strategic Materials." *A Week of the War*, 15 Aug. 1942, 4.
Thomson, A. L. "Exploring the Possibilities of Waste Materials." *Public Cleansing*, Feb. 1940, 184–5.
"From 'May' to 'Must' – Compulsion at Last." *Public Cleansing*, Oct. 1940, 62–3.
"Public Cleansing in National and in a War-Time Economy." *Public Cleansing*, May 1939, 362–70.
"What Is Waste? An Intriguing Question." *Public Cleansing and Salvage*, July 1945, 462–3.
"Tower Guns for Scrap." *Waste Trade World and the Iron and Steel Scrap Review*, 24 Aug. 1940, 2.
"The Trade in Parliament." *Waste Trade World and the Iron and Steel Scrap Review*, 12 July 1941, 6.
"Transactions in Copper." *Waste Trade World and the Iron and Steel Scrap Review*, 26 Aug. 1939, 11.

Ungewitter, Claus, ed. *Verwertung des Wertlosen*. Berlin: Wilhelm Limpert, 1938. *Science and Salvage*, translated by L. A. Ferney and G. Haim. London: Crosby Lockwood and Son, 1944.
United States. War Production Board. *Get in the Scrap*. Washington: U.S. Government Printing Office, 1942.
"The Village Dump." *Public Cleansing and Salvage*, Feb. 1945, 206.
"Vintage Pumping Engines for Scrap." *Waste Trade World and the Iron and Steel Scrap Review*, 7 June 1941, 5.
"W.V.S. and Salvage." WVS *Bulletin*, June 1946, 6.
"Waste Paper in Front Line of Essential Supplies." *Waste Trade World and the Iron and Steel Scrap Review*, 14 Sept. 1940, 11.
"Waste Paper Merchants' Annual Banquet in London." *Waste Trade World and the Iron and Steel Scrap Review*, 14 Jan. 1939, 10–11.
Waste Paper Recovery Association. *Annual Report, 1955*. London: n.p., 1956.
Webster, Russell. Letter to the editor. *Public Cleansing and Salvage*, July 1944, 460.
"We Have Not Reached This!" *Public Cleansing*, May 1940, 298.
Wells, H. G. *The Work, Wealth and Happiness of Mankind*. London: William Heinemann, 1932.
White, Carl M., and P. S. J. Welsford. "The Inter-Allied Book Centre in London." *Library Quarterly* 16, no. 1 (1946): 57–62.
Wigmore, Ethel. "The War and British Medical Libraries." *Bulletin of the Medical Library Association* 34, no. 3 (1946): 151–66, esp. 151.
Women's Voluntary Services. *What a Cog Should Know*. [London]: Women's Voluntary Services, 1941.
"Wood Pulp from Sweden." *Public Cleansing and Salvage*, July 1945, 463.
Worswick, G. D. N. "British Raw Material Controls." *Oxford Economic Papers* 6 (April 1942): 1–41.
The Raw Material Controls. [London]: Fabian Society, [1942].
Wright, Arthur. Letter to the editor. *Waste Trade World and the Iron and Steel Scrap Review*, 17 Aug. 1940, 8.
Y. Y. [Robert Lynd], "Goodbye to Railings." *New Statesman and Nation*, 14 Dec. 1940, 617–18.

Later Publications

Abramson, Rudy. *Spanning the Century: The Life of W. Averell Harriman, 1891–1986*. New York: W. Morrow, 1992.
Ackerman, Frank, and Kevin Gallagher. "Mixed Signals: Market Incentives, Recycling, and the Price Spike of 1995." *Resources, Conservation & Recycling* 35 (June 2002): 275–95.
Armstrong, Robert. "Maud, John Primatt Redcliffe, Baron Redcliffe-Maud (1906–1982)." In *Oxford Dictionary of National Biography*. Oxford: Oxford University Press, 2004.
Barnett, Correlli. *The Audit of War: The Illusion and Reality of Britain as a Great Nation*. London: Macmillan, 1986.
Bishop, Alan. "Brittain, Vera Mary (1893–1970)." In *Oxford Dictionary of National Biography*. Oxford: Oxford University Press, 2004.

Blake, Robert. "Lindemann, Frederick Alexander, Viscount Cherwell (1886–1957)." In *Oxford Dictionary of National Biography*. Oxford: Oxford University Press, 2004.

Boyce, D. George. "Aitken, William Maxwell, First Baron Beaverbrook (1879–1964)." In *Oxford Dictionary of National Biography*. Oxford: Oxford University Press, 2004.

Calder, Angus. *The People's War: Britain, 1939–1945*. New York: Pantheon, 1969.

Campbell, Thomas M., and George C. Herring. Introduction to *The Diaries of Edward R. Stettinius, Jr., 1943–1946*. New York: New Viewpoints, 1975.

Cantwell, John D. *The Second World War: A Guide to Documents in the Public Record Office*, 3d ed. Kew: Public Record Office, 1998.

Claasen, Adam. "Blood and Iron, and der Geist des Atlantiks: Assessing Hitler's Decision to Invade Norway." *Journal of Strategic Studies* 20, no. 3 (Sept. 1997): 71–96.

Clark, J. F. M. "Dawes, Jesse Cooper (1878–1955)." In *Oxford Dictionary of National Biography*. Oxford: Oxford University Press, 2004.

Collingham, Lizzie. *The Taste of War: World War II and the Battle for Food*. New York: Penguin, 2012.

Conquest, Robert. *The Great Terror: A Reassessment*. New York: Oxford University Press, 2007.

Cooke, Miriam. "War, Gender, and Military Studies." *NWSA Journal* 13, no. 3 (Autumn 2001): 181–8.

Cooper, Tim. "Burying the 'Refuse Revolution': The Rise of Controlled Tipping in Britain, 1920–1960." *Environment and Planning A* 42, no. 5 (May 2010): 1033–48.

"Challenging the 'Refuse Revolution': War, Waste and the Rediscovery of Recycling, 1900–50." *Historical Research* 81, no. 214 (Nov. 2008): 710–31.

"Modernity and the Politics of Waste in Britain." In *Nature's End: History and the Environment*, edited by Sverker Sörlin and Paul Warde, 247–72. London: Palgrave Macmillan, 2009.

"War on Waste? The Politics of Waste and Recycling in Post-War Britain, 1950–1975." *Capitalism Nature Socialism* 20, no. 4 (Dec. 2009): 53–72.

Cull, Nicholas John. *Selling War: The British Propaganda Campaign against American "Neutrality" in World War II*. New York: Oxford University Press, 1995.

Davies, Norman. *No Simple Victory: World War II in Europe, 1939–1945*. New York: Penguin, 2007.

de Grazia, Victoria. *How Fascism Ruled Women: Italy, 1922–1945*. Berkeley: University of California Press, 1992.

de Normann, Eric. "Hicks, (Ernest) George (1879–1954)." Rev. Marc Brodie. In *Oxford Dictionary of National Biography*. Oxford: Oxford University Press, 2004.

Denton, Chad. "'Récupérez!' The German Origins of French Wartime Salvage Drives, 1939–1945." *Contemporary European History* 22, no. 3 (Aug. 2013): 399–430.

Desrochers, Pierre. "Does the Invisible Hand Have a Green Thumb? Incentives, Linkages, and the Creation of Wealth out of Industrial Waste in Victorian England." *Geographical Journal* 175, no. 1 (March 2009): 3–16.

Dutton, D. J. "Wood, Edward Frederick Lindley, First Earl of Halifax (1881–1959)." In *Oxford Dictionary of National Biography*. Oxford: Oxford University Press, 2004.

Edgerton, David. *Britain's War Machine: Weapons, Resources, and Experts in the Second World War*. Oxford: Oxford University Press, 2011.

Eksteins, Modris. *Rites of Spring: The Great War and the Birth of the Modern Age*. Boston: Houghton Mifflin, 1989.

Ellis, Roger H. "Recollections of Sir Hilary Jenkinson." *Journal of the Society of Archivists* 4, no. 4 (1971): 261–75.

Enloe, Cynthia H. *Maneuvers: The International Politics of Militarizing Women's Lives*. Berkeley: University of California Press, 2000.

Evans, Bryce. *Ireland during the Second World War: Farewell to Plato's Cave*. Manchester: Manchester University Press, 2014.

Field, Geoffrey G. *Blood, Sweat, and Toil: Remaking the British Working Class, 1939–1945*. Oxford: Oxford University Press, 2011.

Floud, Roderick, and Deirdre McCloskey. *The Economic History of Britain since 1700, Vol. 3: 1939–1992*, 2d ed. New York: Cambridge University Press, 1994.

Fox, Celina. "The Battle of the Railings." *AA Files* 29 (Summer 1995): 50–60.

Foyle's War. "War Games." 3 Nov. 2003.

Friedrich, Jörg. *The Fire: The Bombing of Germany, 1940–1945*. Translated by Allison Brown. New York: Columbia University Press, 2006.

Fry, Geoffrey K. "Fisher, Sir (Norman Fenwick) Warren (1879–1948)." In *Oxford Dictionary of National Biography*. Oxford: Oxford University Press, 2004.

Gillespie, Gordon. "Waring [*married name* Harnett], Dorothy Grace [*pseud.* D. Gainsborough Waring] (1891–1977)." In *Oxford Dictionary of National Biography*. Oxford: Oxford University Press, 2004.

Goodwin, Nathan Dylan. *Hastings at War, 1939–1945*. Chichester: Phillimore, 2005.

Gordon, Peter. "Wake, Joan (1884–1974)." In *Oxford Dictionary of National Biography*. Oxford: Oxford University Press, 2004.

Graham, Virginia. *The Story of WVS*. London: HMSO, 1959.

Graves, Charles. *Women in Green: The Story of the W.V.S.* London: William Heinemann, 1948.

Great Britain. *The Introduction of the Ban on Swill Feeding: 1st Report*. London: TSO, 2007.

Green, E. H. H. "Law, Andrew Bonar (1858–1923)." In *Oxford Dictionary of National Biography*. Oxford: Oxford University Press, 2004.

Grieves, Keith. "Duncan, Sir Andrew Rae (1884–1952)." In *Oxford Dictionary of National Biography*. Oxford: Oxford University Press, 2004.

"Stanley, Edward George Villiers, Seventeenth Earl of Derby (1865–1948)." In *Oxford Dictionary of National Biography*. Oxford: Oxford University Press, 2004.

Griffiths, Richard. "Russell, Hastings William Sackville, Twelfth Duke of Bedford (1888–1953)." In *Oxford Dictionary of National Biography*. Oxford: Oxford University Press, 2004.

Gullace, Nicoletta. *"The Blood of Our Sons": Men, Women and the Renegotiation of British Citizenship during the Great War.* New York: Palgrave Macmillan, 2004.
Hague, Arnold. *The Allied Convoy System, 1939–1945: Its Organization, Defence and Operation.* Annapolis, Md.: Naval Institute Press, 2000.
Harbutt, Fraser J. *The Iron Curtain: Churchill, America, and the Origins of the Cold War.* Oxford: Oxford University Press, 1986.
Herring, George C. *Aid to Russia, 1941–1946: Strategy, Diplomacy, the Origins of the Cold War.* New York: Columbia University Press, 1973.
Hinsley, F. H., et al. *British Intelligence in the Second World War: Its Influence on Strategy and Operations.* New York: Cambridge University Press, 1979.
Hinton, James. *Women, Social Leadership, and the Second World War: Continuities of Class.* Oxford: Oxford University Press, 2002.
Hohn, Uta. "The Bomber's Baedeker: Target Book for Strategic Bombing in the Economic Warfare against German Towns, 1943–45." *GeoJournal* 34, no. 2 (Oct. 1994): 213–30.
Holman, Valerie. *Print for Victory: Book Publishing in England, 1939–1945.* London: British Library, 2008.
Hounsell, Peter. *London's Rubbish: Two Centuries of Dirt, Dust and Disease in the Metropolis.* Stroud: Amberley, 2013.
Howell, David. "Morrison, Herbert Stanley, Baron Morrison of Lambeth (1888–1965)." In *Oxford Dictionary of National Biography.* Oxford: Oxford University Press, 2004.
Howlett, Peter. *Fighting with Figures: A Statistical Digest of the Second World War.* London: Central Statistical Office, 1995.
Hurstfield, Joel. *The Control of Raw Materials.* History of the Second World War. United Kingdom Civil Series. London: HMSO, 1953.
Jones, Helen. "Dalton, (Florence) Ruth, Lady Dalton (1890–1966)." In *Oxford Dictionary of National Biography.* Oxford: Oxford University Press, 2004.
Joyce, Patrick. *The Rule of Freedom: Liberalism and the Modern City.* London: Verso, 2003.
Kandiah, Michael D. "Marquis, Frederick James, First Earl of Woolton (1883–1964)." In *Oxford Dictionary of National Biography.* Oxford: Oxford University Press, 2004.
Karlsgodt, Elizabeth Campbell. "Recycling French Heroes: The Destruction of Bronze Statues under the Vichy Regime." *French Historical Studies* 29, no. 1 (Winter 2006): 143–81.
Kimball, Warren F. *The Most Unsordid Act: Lend-Lease, 1939–1941.* Baltimore: Johns Hopkins University Press, 1969.
Kindleberger, Charles P. *A Financial History of Western Europe*, 2d ed. New York: Oxford University Press, 1993.
King, David. *The Commissar Vanishes: The Falsification of Photographs and Art in Stalin's Russia.* New York: Metropolitan Books, 1997.
Kohan, Charles Mendel. *Works and Buildings.* History of the Second World War. United Kingdom Civil Series. London: HMSO, 1952.
Koss, Stephen. *The Rise and Fall of the Political Press in Britain, Vol. 2: The Twentieth Century.* Chapel Hill: University of North Carolina Press, 1984.

Lavery, Brian. *Churchill Goes to War: Winston's Wartime Journeys*. London: Naval Institute Press, 2007.

Lever, Jeremy. "Greene, Wilfrid Arthur, Baron Greene (1883–1952). In *Oxford Dictionary of National Biography*. Oxford: Oxford University Press, 2004.

Longmate, Norman. *How We Lived Then: A History of Everyday Life during the Second World War*. London: Hutchinson, 1971.

Luckin, Bill. "Pollution in the City." In *The Cambridge Urban History of Britain, Vol. 3: 1840–1950*, edited by Martin Daunton, 207–28. Cambridge: Cambridge University Press, 2000.

"MV Storsten." Clydebuilt Ships Database. www.clydesite.co.uk/clydebuilt/viewship.asp?id=3956, accessed 29 July 2013.

Marwick, Arthur. *Britain in the Century of Total War: War, Peace and Social Change, 1900–1967*. Boston: Little, Brown, 1968.

McIntyre, Ian. "Reith, John Charles Walsham, First Baron Reith (1889–1971)." In *Oxford Dictionary of National Biography*. Oxford: Oxford University Press, 2004.

McNeill, J. R. *Something New under the Sun: An Environmental History of the Twentieth-Century World*. New York: W. W. Norton, 2000.

Melosi, Martin V. *Garbage in the Cities: Refuse, Reform, and the Environment, 1880–1980*. College Station: Texas A&M University Press, 1981.

The Sanitary City: Urban Infrastructure in America from Colonial Times to the Present. Baltimore: Johns Hopkins University Press, 2000.

Moon, Antonia. "Destroying Records, Keeping Records: Some Practices at the East India Company and at the India Office." *Archives* 33, no. 119 (Oct. 2008): 114–25.

Morgan, Kenneth O. "Addison, Christopher, First Viscount Addison (1869–1951)." In *Oxford Dictionary of National Biography*. Oxford: Oxford University Press, 2004.

"George, Lady Megan Arfon Lloyd (1902–1966)." In *Oxford Dictionary of National Biography*. Oxford: Oxford University Press, 2004.

Overy, Richard. *The Twilight Years: The Paradox of Britain between the Wars*. New York: Viking Penguin, 2009.

Padfield, Peter. *War beneath the Sea: Submarine Conflict during World War II*. New York: John Wiley, 1998.

Peiso, O. O. et al. "A Review of Exotic Animal Disease in Great Britain and in Scotland Specifically between 1938 and 2007." *PLoS ONE* 6, no. 7 (July 2011): e22066.

Perrow, Charles B. *Normal Accidents: Living with High Risk Technologies*, rev. ed. Princeton, NJ: Princeton University Press, 1999.

Pimlott, Ben. "Wilmot, John Charles, Baron Wilmot of Selmeston (1893–1964)." In *Oxford Dictionary of National Biography*. Oxford: Oxford University Press, 2004.

Potter, Simon J. "Baillieu, Clive Latham, First Baron Baillieu (1889–1967)." In *Oxford Dictionary of National Biography*. Oxford: Oxford University Press, 2004.

Price, Leah. *How to Do Things with Books in Victorian Britain*. Princeton, N.J.: Princeton University Press, 2012.

Proctor, Tammy M. "'Patriotism Is Not Enough': Women, Citizenship, and the First World War." *Journal of Women's History* 17, no. 2 (Summer 2005): 169–76.

Ridley, Rosalind M., and Harry F. Baker. *Fatal Protein: The Story of CJD, BSE, and Other Prion Diseases*. New York: Oxford University Press, 1998.

Riley, Mark. "From Salvage to Recycling: New Agendas or Same Old Rubbish?" *Area* 40, no. 1 (March 2008): 79–89.

Rockoff, Hugh T. "Keep on Scrapping: The Salvage Drives of World War II." Sept. 2007. NBER Working Paper Series, vol. w13418, 2007, http://ssrn.com/abstract=1014795, accessed 17 Dec. 2012.

Rohrer, Finlo. "What's a Little Debt between Friends?" *BBC News Magazine*, 10 May 2006, http://news.bbc.co.uk/2/hi/uk_news/magazine/4757181.stm, accessed 29 Oct. 2014.

Rohwer, Jürgen. *Axis Submarine Successes of World War Two: German, Italian, and Japanese Submarine Successes, 1939–1945*. Annapolis, Md.: Naval Institute Press, 1999.

Roll, David L. *The Hopkins Touch: Harry Hopkins and the Forging of the Alliance to Defeat Hitler*. New York: Oxford University Press, 2013.

Room, Adrian. "Harrap, George Godfrey (1868–1938)." In *Oxford Dictionary of National Biography*. Oxford: Oxford University Press, 2004.

Russell, Edmund. *War and Nature: Fighting Humans and Insects with Chemicals from World War I to Silent Spring*. New York: Cambridge University Press, 2001.

Scanlon, John. *On Garbage*. London: Reaktion Books, 2005.

Scott, James C. *Seeing Like a State: How Certain Schemes to Improve the Human Condition Have Failed*. New Haven, Conn.: Yale University Press, 1999.

Scott, K. "Window of Deceit." *Army Quarterly & Defence Journal* 123, no. 1 (Jan. 1993): 39–42.

Sheffield, J. V. "Portal, Wyndham Raymond, Viscount Portal (1885–1949)." In *Oxford Dictionary of National Biography*. Oxford: Oxford University Press, 2004.

Sherwood, Robert E. *Roosevelt and Hopkins: An Intimate History*. New York: Harper & Brothers, 1948.

Stamp, Gavin. "Scott, Sir Giles Gilbert (1880–1960)." In *Oxford Dictionary of National Biography*. Oxford: Oxford University Press, 2004.

Stokes, Raymond G., Roman Köster, and Stephen Sambrook. *The Business of Waste: Great Britain and Germany, 1945 to the Present*. New York: Cambridge University Press, 2013.

Stone, Tessa. "Denman, Gertrude Mary, Lady Denman (1884–1954)." In *Oxford Dictionary of National Biography*. Oxford: Oxford University Press, 2004.

Strasser, Susan. *Waste and Want: A Social History of Trash*. New York: Metropolitan Books, 1999.

Summers, Julie. "Farrer, Dame Frances Margaret (1895–1977)." In *Oxford Dictionary of National Biography*. Oxford: Oxford University Press, 2004.

Suri, Jeremi. *Henry Kissinger and the American Century*. Cambridge, Mass.: Belknap Press of Harvard University Press, 2007.

Thorsheim, Peter. "Green Space and Class in Imperial London." In *The Nature of Cities: Culture, Landscape, and Urban Space*, edited by Andrew C. Isenberg, 24–37. Rochester: University of Rochester Press, 2006.

"Salvage and Destruction: The Recycling of Books and Manuscripts in Great Britain during the Second World War." *Contemporary European History* 22, no. 3 (Aug. 2013): 431–52.

Tooze, Adam. *The Wages of Destruction: The Making and Breaking of the Nazi Economy*. New York: Viking, 2006.

Vickers, Hugo. "Hardinge, Alexander Henry Louis, Second Baron Hardinge of Penhurst (1894–1960)." In *Oxford Dictionary of National Biography*. Oxford: Oxford University Press, 2004.

Weaver, Michael. "International Cooperation and Bureaucratic In-Fighting: American and British Economic Intelligence Sharing and the Strategic Bombing of Germany, 1939–41." *Intelligence & National Security* 23, no. 2 (April 2008): 153–75.

Weber, Heike. "Towards 'Total' Recycling: Women, Waste and Food Waste Recovery in Germany, 1914–1939." *Contemporary European History* 22, no. 3 (Aug. 2013): 371–97.

Wilford, R. A. "Parker, Dame Dehra (1882–1963)." In *Oxford Dictionary of National Biography*. Oxford: Oxford University Press, 2004.

Windlesham. "Isaacs, Stella, Marchioness of Reading and Baroness Swanborough (1894–1971)." In *Oxford Dictionary of National Biography*. Oxford: Oxford University Press, 2004.

Winter, James. *London's Teeming Streets, 1830–1914*. London: Routledge, 1993.

Women's Voluntary Service for Civil Defence. *Report on 25 Years Work, 1938–1963*. London: HMSO, 1963.

Woodward, Donald. "Swords into Ploughshares: Recycling in Pre-Industrial England." *Economic History Review* 38, no. 2 (May 1985): 175–91.

Zimring, Carl A. *Cash for Your Trash: Scrap Recycling in America*. New Brunswick, N.J.: Rutgers University Press, 2005.

Zweiniger-Bargielowska, Ina. *Austerity in Britain: Rationing, Controls, and Consumption, 1939–1955*. Oxford: Oxford University Press, 2000.

Index

Other Books in the Series *(Continued from page iii)*

Mark Elvin and Tsui'jung Liu *Sediments of Time: Environment and Society in Chinese History*

Richard H. Grove *Green Imperialism: Colonial Expansion, Tropical Island Edens and the Origins of Environmentalism, 1600–1860*

Elinor G. K. Melville *A Plague of Sheep: Environmental Consequences of the Conquest of Mexico*

J. R. McNeill *The Mountains of the Mediterranean World: An Environmental History*

Theodore Steinberg *Nature Incorporated: Industrialization and the Waters of New England*

Timothy Silver *A New Face on the Countryside: Indians, Colonists, and Slaves in the South Atlantic Forests, 1500–1800*

Michael Williams *Americans and Their Forests: A Historical Geography*

Donald Worster *The Ends of the Earth: Perspectives on Modern Environmental History*

Samuel P. Hays *Beauty, Health, and Permanence: Environmental Politics in the United States, 1955–1985*

Warren Dean *Brazil and the Struggle for Rubber: A Study in Environmental History*

Robert Harms *Games against Nature: An Eco-Cultural History of the Nunu of Equatorial Africa*

Arthur F. McEvoy *The Fisherman's Problem: Ecology and Law in the California Fisheries, 1850–1980*

Alfred W. Crosby *Ecological Imperialism: The Biological Expansion of Europe, 900–1900*, Second Edition

Kenneth F. Kiple *The Caribbean Slave: A Biological History*

Donald Worster *Nature's Economy: A History of Ecological Ideas*, Second Edition

Made in the USA
San Bernardino, CA
19 November 2017